ZEW Economic Studies

Publication Series of the Centre for European Economic Research (ZEW), Mannheim, Germany

ZEW Economic Studies

Further volumes of this series can be found at our homepage:
springer.com/series/4041

Vol. 16: U. Kaiser
Innovation, Employment, and Firm Performance in the German Service Sector
2002. X, 164 pp. ISBN 3-7908-1481-4

Vol. 17: J. Köke
Corporate Governance in Germany
An Empirical Investigation
2002. VIII, 184 pp. ISBN 3-7908-1511-X

Vol. 18: E. Wolf
What Hampers Part-Time Work?
An Empirical Analysis of Wages, Hours Restrictions and Employment from a Dutch-German Perspective
2003. X, 174 pp. ISBN 3-7908-0006-6

Vol. 19: P. Cecchini et al. (Eds.)
The Incomplete European Market for Financial Services
2003. XII, 255 pp. ISBN 3-7908-0013-9

Vol. 20: C. Böhringer, A. Löschel (Eds.)
Empirical Modeling of the Economy and the Environment
2003. VI, 326 pp. ISBN 3-7908-0078-3

Vol. 21: K. Rennings, T. Zwick (Eds.)
Employment Impacts of Cleaner Production
2003. VIII, 322 pp. ISBN 3-7908-0093-7

Vol. 22: O. Bürgel, A. Fier, G. Licht, G. Murray
The Internationalisation of Young High-Tech Firms
An Empirical Analysis in Germany and the United Kingdom
2003. XII, 291 pp. ISBN 3-7908-0120-8

Vol. 23: F. Hüfner
Foreign Exchange Intervention as a Monetary Policy Instrument
Evidence for Inflation Targeting Countries
2004. X, 175 pp. ISBN 3-7908-0128-3

Vol. 24: E. Lüders
Economic Foundation of Asset Price Processes
2004. XII, 121 pp. ISBN 3-7908-0149-6

Vol. 25: F. Reize
Leaving Unemployment for Self-Employment
An Empirical Study
2004. XII, 241 pp. ISBN 3-7908-0168-2

Vol. 26: Ch. Böhringer, A. Löschel (Eds.)
Climate Change Policy and Global Trade
2004. VIII, 381 pp. ISBN 3-7908-0171-2

Vol. 27: K. Jacob et al.
Lead Markets for Environmental Innovations
2005. XII, 273 pp. ISBN 3-7908-0164-X

Vol. 28: L. Lammersen, R. Schwager
The Effective Tax Burden of Companies in European Regions
2005. XI, 251 pp. ISBN 3-7908-1562-4

Vol. 29: Ch. Elschner, R. Schwager
The Effective Tax Burden on Highly Qualified Employees
2005. VIII, 123 pp. ISBN 3-7908-1568-3

Vol. 30: Ch. Lauer
Education and Labour Market Outcomes
2005. IX, 286 pp. ISBN 3-7908-1569-1

Vol. 31: Ch. Böhringer, A. Lange (Eds.)
Applied Research in Environmental Economics
2005. VI, 314 pp. ISBN 3-7908-1587-X

Vol. 32: O. Heneric, G. Licht, W. Sofka (Eds.)
Europe's Automotive Industry on the Move
Competitiveness in a Changing World
2005. XI, 275 pp. ISBN 3-7908-1590-X

Vol. 33: Th. Hempell
Computers and Productivity
How Firms Make a General Purpose Technology Work
2006. X, 194 pp. ISBN 3-7908-1647-7

Vol. 34: D. Stegarescu
Decentralised Government in a Integrating World
Quantitative Studies for OECD Countries
2006. XI, 231 pp. ISBN 3-7908-1668-8

Matthias Meitner

The Market Approach to Comparable Company Valuation

With 34 Figures and 26 Tables

Physica-Verlag
A Springer Company

ZEW
Zentrum für Europäische
Wirtschaftsforschung GmbH
Centre for European
Economic Research

Series Editor
Prof. Dr. Dr. h.c. mult. Wolfgang Franz

Author
Dr. Matthias Meitner
Arcisstraße 58
80799 München
Germany
matthias_meitner@hotmail.com

Diss., Univ. Erlangen-Nürnberg, n2

ISBN-10 3-7908-1722-8 Physica-Verlag Heidelberg New York
ISBN-13 978-3-7908-1722-5 Physica-Verlag Heidelberg New York

Cataloging-in-Publication Data applied for
Library of Congress Control Number: 2006902903

Physica-Verlag is a part of Springer Science+Business Media

springer.com

Printed in Germany

Cover design: Michael Weiland, ZEW, Mannheim

SPIN 11740551 43/3153-5 4 3 2 1 0 – Printed on acid-free paper

"Value is a relative term. The value of a thing means the quantity of some other thing, or of things in general, which it exchanges for."

— John Stuart Mill

Principles of Political Economy, 1848

"Bewerten heißt vergleichen"

— Adolf Moxter

Grundsätze ordnungsmäßiger Unternehmensbewertung, 2nd edition, 1983

Acknowledgements

First and foremost I am indebted to my supervisor, Prof. Wolfgang Gerke, who motivated me and energised this work from start to finish. His wit, vast experience and captivating style of teaching stirred in me an early fascination and enthusiasm for finance when I was a student at Erlangen-Nürnberg. I am grateful to my co-supervisor, Prof. Volker H. Peemöller, for his kindness and encouragement. Furthermore, I would like to thank Prof. Wolfgang Franz, President of the Centre for European Economic Research (ZEW), for supporting this dissertation. The many discussions with my colleagues at the ZEW saved this work from turning into an empirical disaster. Special thanks go to Prof. François Laisney, Marcus Kappler, Dr. Michael Schröder, Wolfgang Sofka (the "matching-expert") and Dr. Peter Westerheide. I also thank Prof. Richard Deaves who made my two months visiting scholarship at McMaster University in Hamilton, Ontario/Canada a complete success. The many helpful comments from Carolin Amann, Christoph Beckmann, Dr. Felix Hüfner and Andreas Kucher are highly appreciated. Kelly Holzscheiter, Jan Hoon, Waldemar Rotfuß, Christian Schwab and Elena Todorova provided excellent research assistance. My thanks also go to the Global Competency Centre of PricewaterhouseCoopers for providing grant assistance to this work. Finally, I want to express my deep gratitude to my dear mother for supporting me during my studies and for believing in me.

Mannheim, Germany,
February 2006

Matthias Meitner

Table of Contents

1 Introduction

1.1 Motivation

Company valuation is one of the most important tasks of financial analysts, investors, consultants, and managers. It not only provides the basis for their decision to purchase or sell whole companies or shares of a company. It is also indispensable for the application of a sound value based management and successful restructuring. However, the process of valuing a company is complex and not standardised at all. There are many different interpretations of what "value" means, and there are many different approaches to determine this value. The valuation approach that enjoys the most widespread popularity in theory and practice is the direct valuation approach, which is based on the net present value concept. The discounted cash flow method is an example for this approach. However, in order to better deal with project flexibility it is sometimes proposed to apply a real options approach. This approach shares high reputation amongst theoreticians and is subject to a vast range of academic papers, but so far it is of almost no importance in valuation practice. The direct opposite of real options valuation – in terms of popularity amongst academics and practitioners – is the relative valuation approach. While this approach is of paramount relevance in real world valuations, literature generally dislikes it and calls it a "quick and dirty method of valuation" (Benninga and Sarig, 1997: 330) that lacks theoretical foundation. Comparable company valuation is a variant of relative valuation. It is based on the principle of arbitrage and values companies based on how other, similar companies are valued. If these similar companies are publicly listed, then the valuation method is called *the market approach to comparable company valuation.*

The wide recognition of the market approach to comparable company valuation amongst practitioners has three causes. First, it is easy to use. In fact, once comparable companies and the valuation model are chosen, the application is straightforward and does not require any specific skills. Second, comparable company valuation relies on existing market prices of companies. Therefore, no explicit forecasts of the cash flow development of the valuation objective are necessary. Moreover, comparable company valuation better reflects the current mood of the market than direct valuation approaches. Third, a relative valuation is easier to present to clients and customers than direct valuations.

In contrast, there are also three crucial reasons for the lack of academic acceptation of comparable company valuation. The first reason is a purely technical one. Comparable company valuation requires certain valuation circumstances that direct valuation approaches do not (necessarily) require. In particular, these circumstances are a set of companies that are similar to the valuation object and a functioning market that fairly prices these comparable companies. In this context, opponents of the comparable company valuation approach point out that the stock market is far from being perfectly efficient and that it is hardly possible to find two identical companies (not to mention the problem of finding more than two equal companies). The second reason is rather ideological in nature. Comparable company valuation is often accused of being a static investment approach that does not conform to basic valuation principles because of a lack of future orientation.[1] The third reason concerns the concrete application of comparable company valuation models. Because of the trade-off between easy-to-handle valuation models and the difficulty of properly determining the input factors, comparable company valuation risks suffering from a "garbage in – garbage out" problem. To put it more precisely, comparable company valuation models can be easily used but even more so, easily misused (see e.g. Damodaran, 2002: 453).

These two different attitudes make comparable company valuation one of the most controversial valuation approaches. While conflicting standpoints of theory and practice are nothing unusual in finance[2], it seems that – with regard to the attempt to bridge the gap between these two positions – the potential is not tapped to its fullest extent here. In fact, most theoretical research sticks to formal discussions. Valuation models are typically judged by the plausibility of their assumptions, not by their ability to accurately value companies. One of the biggest problems in this context is that the forecasting challenges – which are inherent in every valuation approach – are often suppressed in the discussions.[3] Consequently, still little is known about the differences of forecasting requirements between different valuation approaches and how forecasting problems can be reduced. As a consequence, most theoretical research is limited in terms of its usefulness to investors since it cannot serve as a guideline in valuation practice (see also Born, 1995: 7-9; Bernard, 1989: 87-91). The empirical literature does not add much to reduce this discrepancy, either. Of course, recently some studies have well contributed to a better understanding of how comparable company valuation functions (see e.g. Herrmann, 2002; Richter and Herrmann, 2002; Liu et al., 2002; Bhojraj and Lee, 2002; Baker and Ruback, 1999; Beatty et al., 1999). However, their number is few and they rarely render concrete advice for how to deal with real world valuation problems.

What is especially noticeable is the lack of differentiating research (both theoretical and empirical), i.e. research that considers that valuation models cannot reasonably be applied for every company and in every valuation situation, or re-

---

1 For a list of academic criticism of comparable company valuation, see Peemöller et al. (2002: 199-201).

2 Just think about the severe theoretical criticism of the Capital Asset Pricing Model (see e.g. Hering 2003: 283-296) which could not prevent that this model is by far the most popular tool to determine the cost of equity in real world direct valuations.

3 A good forecast is at least as important as a reasonable valuation model. Lee (1999: 414) states in this context that the "essential task in valuation is forecasting. It is the forecast that breathes life into a valuation model".

search that analyses which valuation model is best to use under certain circumstances.[4] This non-existent situational research is a major obstacle in better understanding the whole comparable company valuation process, and one of the main reasons for practitioners' low acceptance of academic findings.

1.2 Research Aims

The purpose of the research presented here is to contribute to the literature by providing a systematic study on the nature and significance of the *market approach to comparable company valuation* from a German perspective. Due to the variety of unresolved issues in comparable company valuation, this study does not address one big research question but rather several smaller questions. The answers to these questions should – as a whole – help draw a more complete picture of the comparable company valuation process. Light will be shed on comparable company valuation from both a theoretical and empirical perspective. The empirical part consists of three smaller surveys amongst financial analysts and institutional investors, and of a broad econometric study. In spite of the sometimes rather formal proceeding (both in the theoretical and the empirical part) special emphasis is on economic content and usefulness to practitioners. In order to ensure this usefulness to the practice, a differentiated proceeding is sometimes necessary. This especially means that many aspects should be discussed, analysed and empirically tested dependent on different valuation circumstances. By doing this, concrete advice can be given to appraisers on how to behave under these valuation circumstances. It is important to notice that while the theoretical part of this examination concerns all facets of the comparable company valuation process – selection of comparable companies, valuation model choice, application range etc. – the focus of the empirical part is clearly on valuation model choice.

The following five batteries of questions will be addressed in this study:

- How does comparable company valuation fit into the business valuation framework? What is the link to other valuation approaches? What is the application range of comparable company valuation?
- What are the determinants of the two main tasks in comparable company valuation (the selection of comparable companies and the valuation model choice)? How can appraisers interpret the influence of these determining factors? How can they deal with changes in these determinants?
- What are the implications and problems associated with classical single-factor comparable company valuation models (such as the price-earnings ratio)? What forecasts are necessary in order to adequately apply these models?

---

4 Some of the rare examples are provided by Kim and Ritter (1999) who analysed the aptitude of multiples in the pricing of Initial Public Offerings, and Gilson et al. (2000) who examined the valuation of bankrupt firms.

- Can multi-factor models (i.e. models that make use of more than one accounting reference variable) help overcome some of the problems associated with single-factor models? How do the two accounting variables book value of equity and earnings interact in comparable company valuation? What determines the relative valuation roles of book value and earnings?
- What determines the height of the multiples at which companies trade?

1.3 Reading Guide

The study as a whole is divided into six chapters. After the general introduction provided here, *chapter 2* presents the foundations of comparable company valuation, and discusses how this approach fits into the business valuation framework. Part of this chapter is an overview of different value theories, the relationship between the terms "value" and "market price", the links between comparable company valuation and other valuation approaches, as well as the application range of comparable company valuation. Additionally, some special issues in comparable company valuation – such as the requirements concerning the quality of accounting variables, the aggregation of valuation ratios and the use of premiums and discounts are discussed.

Chapter 3 provides a detailed analysis of the two main tasks in comparable company valuation – the selection of comparable companies and the valuation model choice – as well as of their determinants. It is shown that the degree of similarity of comparable companies and the degree of capital market efficiency crucially impact the comparable company selection process. Likewise, the value relevance of the accounting reference variables, the future similarity of companies and potential technical restrictions of valuation models are presented as determinants of the valuation model choice. The chapter closes with an explanation of why comparable company valuation should be understood as an integrated process in which all tasks must interdigitate.

Chapter 4 describes the comparable company valuation process for three kinds of models: immediate, single-factor and multi-factor models. The chapter also contains a presentation of common mistakes in the use of comparable company valuation and of the major shortcomings and problems associated with single-factor models. The main emphasis of this chapter is on the derivation of a two-factor comparable company valuation model based on book value of equity and earnings, which aims at overcoming some of the problems associated with single-factor models. Most contents of chapter 4 – along with some parts of chapter 3 – are based on the author's research paper *How Fundamentals Drive the Equity Value* (Meitner, 2004).

Chapter 5 covers the empirical examinations of value relevance and pricing accuracy. The value relevance study analyses the appropriateness of different valuation models under the assumption that there is no lack of perfectly comparable companies. In addition to some well known econometric models, an innovative approach called the *matching estimator* is applied in this analysis in order to overcome a selection bias problem. This approach originally comes from labour market research, and has not often appeared in financial research literature before. The pricing accuracy study investigates the historical performance of the two-factor model that was derived in chapter 4 as compared to some single-factor and simple multi-factor models.

Finally, some concluding remarks are formulated in *chapter 6*. This chapter summarises the most important findings with regard to comparable company valuation, and contains implications for future research.

2 Foundations of Comparable Company Valuation

2.1 Definitions and Scope

The objective of business valuation is to assign a value to a company. In this context, the term "value" should be understood as the degree of utility that a (potential) investor gains from owning a company (see Muenstermann, 1970: 11; Moxter, 1983: 128; Seppelfricke, 2003: 1). The company for which the corporate value is determined is known as the target company or simply the target. One thing all types of business valuation have in common is that they are performed from the perspective of one of two typical sides of a transaction: the buy-side and the sell-side. From the (potential) buyer's perspective, the value of a company can then be seen as the upper limit of his readiness to pay for that company. From the (potential) seller's perspective, the value of a company can be seen as the lower limit of what he wants to get for that company.[5] Consequently, the process of valuing a company is also to determine potential prices for a company (see Peemöller, 2005a: 3).

Business valuation is not restricted to determining the value of a whole company. It is also a reasonable tool to value an interest in (i.e. shares of) a company. Regarding the scale of corporate assets that can be valued, there are basically two types of company valuation. First, "enterprise valuation" denotes the process of valuing a company as a whole, i.e. to determine the value that belongs to all capital providers. Second, "equity valuation" characterises the process of valuing the part of a company that belongs only to the shareholders. The equity value of a company can be directly calculated by focusing on value components that are relevant only for owners of the company, or indirectly by subtracting the value of non-equity capital from the enterprise value of the company.

To accurately perform company valuation, appraisers have to comply with certain basic requirements (see Peemöller, 2005a: 3):

- Valuation must be future-oriented: Only benefits that will be earned in the future are value relevant (see Muenstermann, 1970: 21).
- Provision for all components that affect utility: Valuation should not be restricted to financial goals; everything that raises utility should be taken into account.

---

5 See Casey (2000: 2) the terms "buyer" and "seller" include those market participants that are not actively involved in transactions but that benefit from the buyer or the seller, respectively.

- Provision for uncertainty about the future: Forecasting models should adequately consider chances (upside potential), but also dangers (downside risk) concerning the future development of corporate profits.
- Investor orientation: The appraiser has to take into account for whom and in which situation the valuation should be performed (see Moxter, 1983: 23-32).

Financial theory uses several techniques to determine corporate values that widely conform with these requirements. Along with those approaches that are based on net present value models (also known as direct valuation approaches) and those that are built on option pricing theory (also known as contingent claim valuation approaches) literature names relative valuation – and especially *comparable company valuation* – as the third general approach.[6] This latter approach is based on the principal of arbitrage that says that all substitutes should sell for the same price (see Gerke and Bank, 2003: 270-271). Thus, the comparable company valuation (CCV) approach values target companies based on how investors value similar companies.

From a methodical perspective, CCV can be divided into three different variants: immediate CCV, single-factor CCV and multi-factor CCV. Immediate CCV describes the process of assigning a value to a target company based on perfect substitutes. Due to the scarcity of totally equal or almost equal companies this approach has little relevance in practical valuation settings.

Single-factor CCV has significantly lower requirements concerning the similarity of the comparable companies because it uses a linking factor that settles minor differences between the comparable companies and the target company. The single-factor approach proceeds in two steps: In the first step the value of a comparable company or the average value of a set of comparable companies has to be expressed as a multiple of a certain – mostly accounting based – basis of reference (such as earnings, EBITDA, sales, etc.) in which the companies differ. In the second step this derived multiple is applied on the respective basis of reference of the target company. This approach – also known as valuation using multiples – covers the most widely used CCV models.

Multi-factor CCV resembles the single-factor approach in that it makes use of linking factors. The only difference is that multi-factor CCV is built on more than one linking factor and therefore on more than one basis of reference. Such multi-factor CCV models can sometimes be found in equity research reports. However, there is no widespread use of this approach in practice.

This work focuses on the most dominant approach to CCV: *the market approach*. The market approach (sometimes also called *similar public company method*) is characterised by the reliance on a set of stock exchange listed comparable companies. One reason why the theoretical and the empirical part both strongly focus on that market approach is better data availability for stock listed

---

6 See Bhojraj and Lee (2002: 413-414); Damodaran (2002: 11). Especially in German literature asset-based valuation is seen as the fourth general valuation approach. This approach is, however, only used under certain valuation circumstances, see section 2.2.3.4.

companies. Still, many of the discussions in the theoretical part and many of the results drawn from the empirical study may also be true for private non-listed firms. It is important to note that this study does not primarily focus on the accounting or taxation aspects associated with CCV but rather on the economic aspects.

2.2 Comparable Company Valuation Within the Business Valuation Framework

CCV is often denoted as a simplified valuation approach (see Seppelfricke, 1999: 301; IDW, 2000: 840; Behringer, 2002: 149; Wiehle et al., 2004: 42; Coenenberg and Schultze, 2002: 700), a "quick and dirty method of valuation" (Benninga and Sarig, 1997: 330) which is not applicable when determining the intrinsic value of a company (see Ballwieser, 1991: 58-60; Buchner and Englert, 1994: 1580; Ballwieser, 1997: 188; Olbrich, 2000: 458-459). Additionally, this approach is subject to a considerable amount of academic criticism, which says it goes against the basic principles of business valuation (see Wiehle et al., 2004: 42; Peemöller et al., 2002: 199-201; Bausch, 2000: 452; Ballwieser, 1991: 62; Benninga and Sarig, 1997: 331). Contrary to that, it is also described as one of the most popular methods in valuation practice (see Damodaran, 2002: 453-454; Löhnert and Böckmann, 2005: 406-408; Nelles et al., 2001: 323; Fernandez, 2002; Kames, 2000: 58-60, 100-101; Wichels, 2002: 146, 148; Duerr, 1995: 27; Kusterer, 2003: 99-100; Creutzmann and Deser, 2005: 2-4; Achleitner, 2004, EVCA, 2005: 13-18). Although some recent studies try to explain or even bridge the gap between theory and practice (see e.g. Richter and Herrmann, 2002; Herrmann, 2002; Peemöller et al., 2002; Beckmann et al., 2003; Kaplan and Ruback, 1995; Baker and Ruback, 1999; Kim and Ritter, 1999; Bhorjraj and Lee, 2002; Liu et al., 2002), some unsettled issues remain. Effectively, the mentioned trade-off is to some extent also due to a widespread uncertainty of how CCV fits into the overall business valuation framework.

The following section should therefore give information with regard to the questions of how the results of the CCV approach (i.e. the appraisal value) can be interpreted and what this means for (potential) investors. Furthermore, the classification below should help to better understand the relationship between CCV and other business valuation approaches. Finally, some light will be shed on the practical applicability of relative valuation approaches.

2.2.1 Value Theories

2.2.1.1 Theory of Objective Value

According to the theory of objective value there is only one exclusive corporate value, which holds for all investors. To put it differently, value is purely a function of the company's economic potential but not of investors' preferences (see Mellerowicz, 1952: 12, 59-60; Engels, 1962: 6-8; Muenstermann, 1970: 21-28; Moxter, 1983: 26-27; Künnemann, 1985: 10-25; Peemöller, 2005a: 4-5; Mandl and Rabel, 1997: 6. Jaensch, 1966: 6-8 argues in a similar vein).

Correspondingly, this theory postulates that the upper limit of a (potential) buyer's readiness to pay for a company exactly equals the lower limit of a potential seller's price demand. As a result, the intersection of both positions is not a range of potential prices but only one price – the price that equals the objective value of a company. Thus, under this theory the terms "value" and "price" can be used interchangeably (see Jaensch, 1966: 7; Engels, 1962: 7; Muenstermann, 1970: 12; Peemöller, 2005a: 4).

A major problem is that this theory fails to explain why potential prices vary depending on the valuation circumstances and the type of investor (see Muenstermann, 1970: 12; Peemöller, 2005a: 4). Proponents of this theory claim that this failure is due to a lack of valuation competence of certain investors and the diversity of valuation methods (see Mellerowicz, 1952: 61-62; Jaensch, 1966: 7).

However, there is also no explanation as to why transactions should take place, because neither of the two participants (buyer or seller) in this transaction benefits from it (see Hering, 2000: 441). In the simplest case of a costless company transfer, the respective wealth positions remain unchanged. Even worse, under the more realistic settings of existing transaction costs both participants would effectively lose money.

To defend at least part of the theory of objective value, it must be assumed that there is not only an objective company value but also personal, economic or strategic preferences of investors (e.g. synergies) beyond this objective value, which finally lead to different price expectations (see Kuennemann, 1985: 24-25, 44-52). In this context, one variant of the objective value is of special importance: The so-called "objectified value" which is generally not observable but can serve as a basis of further adjustments.[7] The Institut der Wirtschaftsprüfer in Deutschland e.V. (IDW – German Institute of Certified Public Accountants), an accounting body with voluntary membership, regards the objectified value as the stand-alone value of a company without consideration of planned but not yet implemented future investments or strategy changes (see IDW, 2000: 829-831, 836-837; Peemöller, 2005a: 6). It is important that since the objectified value does not account for potential synergy effects it is typically close to the value from the perspective of a (potential) seller and might therefore differ from the value seen from the perspective of a (potential) buyer.[8] One advantage of this variant is that the objectified

[7] Whether the objectified value can really be seen as a variant of the objective value is subject to many discussions. However, the classification seems to be reasonable here.

[8] See Moxter (1983: 27-28); IDW (2000: 829-830); with regard to listed companies in Germany the main field of applicance of the objectified value is the so-called "squeeze-out procedure", pursuant to sections 327a et seq. of the German Stock Corporation Law (Aktiengesetz). If a majority shareholder holds at least 95% of a company, the squeeze-out procedure permits him to acquire the shares of the minority

value manages to connect the theory of objective value with the theory of subjective value.

Contrary to the German perspective, the Anglo-Saxon valuation theory regards the objectified value as the fair market value – a value that a typical average investor would assign to the company under average circumstances. Thus, consideration of the impact of future investments and strategy changes is not categorially excluded; rather it depends on what the average investor expects (see Pratt et al., 2000: 28-30).

2.2.1.2 Theory of Subjective Value

In contrast to the theory of objective value, the theory of subjective value particularly emphasizes the investors' perspective. According to this theory the company value is not unique but depends on the set of preferences and expectations of an investor (see Peemöller, 2005a: 6-7; IDW, 2000: 831; Moxter; 1983: 138-145; Engels, 1962: 8-10; Jaensch, 1966: 8-17; Künnemann, 1985: 25-29; Bonbright, 1965: 128). Since preferences and expectations are highly subjective and therefore vary between investors, there might be as many different values for one company as there are valuations (see Jaensch, 1966: 9-10). Under this theory the company value is often called the "practical value" or "value in use".

The theory of subjective value is based strongly on the theory of economizing behaviour (expected-utility theory). It states that the basis for the evaluation of risky alternatives is the utility, i.e. the benefit or satisfaction that a decision maker expects from the choice of each of the alternatives. Thus, decision makers do not only focus on the monetary value of the alternatives' pay-offs but also on the expected utilities of these payoffs (see von Neumann and Morgenstern, 1944; Luce and Raiffa, 1957; Moxter, 1983: 138-139; Varian, 1992: 94-108; Gwartney and Stroup, 1997: 11; Binger and Hoffmann, 1998: 511-521). The theory of subjective value is a generalisation of the expected-utility theory adjusted to the business valuation framework. More precisely, the subjective value theory suggests that investors value companies with respect to their personal degree of risk aversion, their personal tax situation and the alternative investments available to them (see Peemöller, 2005a: 6; Moxter, 1983: 23-24).

The large number of possible values for one company along with the fact that there is usually only one market price implies that value and market price differ in the majority of cases. Effectively, the price is the result of a negotiation between buy-side and sell-side market participants based on the respective subjective company values (see Loistl, 1994: 313; Casey, 2000: 4).

While methodically sound, the theory of subjective value has two shortcomings in valuation practice: First, a third party cannot retrace how the appraiser calculated the company value. In fact, the valuation process resembles a black box since many factors that determine the company value are hidden in the subjectivity of the appraiser (see Peemöller, 2005a: 7). Second, following this theory value de-

shareholders for cash compensation. The amount of this cash compensation, in turn, crucially depends on the objectified value of the company.

termination sometimes fails. The reason for this is that the theory of subjective value does not allow successful arbitration between conflicting parties with extremely different expectations (see Mandl and Rabel, 1997: 8).

2.2.1.3 Theory of Functional Value

The theory of functional value provides a different approach to overcome the drawbacks of the theory of objective value. Moreover, it tries to deliver a higher traceability of the valuation process for third parties.[9] The most important aspect of this theory is that it partitions the reasons for company valuation into different functions. Along with some auxiliary functions – such as tax assessment or forming of contracts – there are four main functions (see Peemöller, 2005a: 8; IDW, 2000: 827; Sieben, 1983: 539-542):

(1) Consultancy function[10]

This function provides assistance either for the buy-side or for the sell-side. While the buyer wants to know the upper limit of his readiness to pay, the seller is interested in the lower limit of what he wants to get. Therefore the aim of the consultancy function is to determine marginal prices based on which decisions can be made (*decision values*) (see Peemöller, 2005a: 8; Moxter, 1983: 13-14; Drukarczyk, 2003: 132; Hering, 1999: 3). Consequently, the main task is to establish a best-case scenario for the respective party. In this context it should be noticed that potential synergies and the potential effects of strategy changes are to be considered for buy-side consultancy.

(2) Intermediation function

Given that the marginal prices of both parties are known or can be externally determined, the intermediation function (also called arbitration function) aims to balance the different preferences fairly. To manage this task, arguments of both parties as well as personal estimations of the appraiser should be considered (see Matschke, 1976: 130-361, Drukarczyk, 2003: 132). It is important to note that it is only possible to find such an intermediate value if the marginal price of the buyer exceeds the marginal price of the seller.

(3) Argument function

The principal task of this function is to collect and disclose arguments that support the intention of one of the two parties. Usually, the goal is to either increase the amount a seller can get or to decrease the amount a buyer has to pay. Even if this function is also trying to influence the other side, great emphasis is placed on the accuracy of the valuation process. However, national and international accoun-

---

[9] The theory of functional value emerged from the so-called „Kölner Schule", precursors of this theory were Muenstermann (1970); Jaensch (1966); Engels (1962); Matschke (1976) and Sieben (1983).

[10] The consultancy function is also called "decision function", see Hering (1999: 3).

tancy bodies do not consider this function compatible with accountants' codes of profession.[11]

(4) Neutral referee function

This function – fulfilled by an independent financial expert (see ASA, 2002: 34-35) – is closely connected with the accountancy profession. The IDW considers this function as one of the typical tasks of an accountant.[12] The goal of the neutral referee function is to value a company without any subjective influence, which in practice means to determine the objectified value of the company. However, in Germany this orientation towards the objectified value (in the sense of the IDW) is sometimes criticized because of the disregard for corporate development possibilities due to a change in ownership (see Moxter, 1983: 27-28).

2.2.2 Value Versus Price

2.2.2.1 Nature of Price

Like the price of any asset the price of a company is the amount of money that balances the different interests of the sell-side and the buy-side in a transaction (see Loistl, 1994: 313; O'Hara, 1995: 3; Gwartney and Stroup, 1997: 62-66; Boecking and Nowak, 1999: 170; Schultze, 2003: 16-17). It is determined by the relationship of the marginal prices of each side, the market participants' relative power to negotiate, the negotiation strategy and the influence of third parties (e.g. auditors, consultancies, investment banks) (see Casey, 2000: 4). The market price of a company, i.e. the product of the price at which shares of the company are quoted on the stock exchange multiplied by the number of shares outstanding, however, differs a great deal from the price of a company as the result of bilateral negotiations.[13]

The market price is a function of the decision values of each single investor (see Pratt et al., 2000: 31). It is a function of the market microstructure, market liquidity and informational market efficiency as well (see O'Hara, 1995: 3-6, 215-250; Schwartz, 1993: 397-437; Casey, 2000: 6; Damodaran, 2001b: 141-146). Usually it does not depend on the relative negotiation power on either side, and

---

11 See IDW (2000: 827); Peemöller (2005a: 10-11). Regarding the problems of methodologically justifying the existence of argument values, see Drukarczyk (2003: 134).

12 See IDW (2000: 827); a major difference between the neutral referee function and the intermediation function is that the neutral referee function does not necessary require an intersection between the decision values of the buy-side and the sell-side. This is especially important in dominated valuation settings, see Drukarczyk (2003: 133). For a distinction between dominated and un-dominated valuation settings, see Matschke (1976: 26-39).

13 See Casey (2000: 141-203); for reasons of simplicity it is assumed here that the company is purely financed with equity. In literature, sometimes price and market price are seen as identical; see e.g. Herrmann (2002: 15).

third parties have no major influence on its determination.[14] However, the assumption incorporated into many theoretical models, that all market participants are price takers and therefore have no impact on market prices (see Copeland et al., 2005: 147), does not precisely describe the way in which market prices are determined. In fact, because of minimum tick sizes the order of a single investor might not suffice to *change* the current market price. Nevertheless, this order moves the supply curve (in the case of a sell order) or the demand curve (in the case of a buy order) of the respective stock and therefore effectively *influences* the market price (see Demsetz, 1968: 33-53; O'Hara, 1995: 3-6).

While market microstructure is the system of specific trading mechanisms and its impact on the price formation process, informational market efficiency refers to the degree to which information is reflected in prices. Market liquidity is a measure of how quickly investors can trade at prices that are reasonable for given supply/demand conditions (measure of marketability) and is closely related to market efficiency. Factors affecting market liquidity are the depth of the market (the amount of orders in the close neighborhood of the current market price), the breadth of the market (the volume of the best buy and sell order) and the resiliency (the ability of the market to restore temporarily biased share values due to order imbalances) (see Schwartz, 1993: 127; O'Hara, 1995: 215-250).

2.2.2.2 Relationship Between Value and Price

In German literature the term "market price" is sometimes seen as basically different from the term "corporate value" (see Busse von Colbe, 1957: 10; Herrmann, 2002: 15; a similar opinion is provided by Bausch, 2000: 457 and Hommel and Braun, 2002: 10-17), while in anglo-american, rather capital market-oriented literature, these two terms are often seen as broadly similar (see e.g. Arrow, 1964: 91-96; Sharpe, 1964: 425-442; Olbrich, 2000: 458; FASB, 2002: 301; a more critical view is provided by Shleifer and Vishny, 1997: 35-55). However, a differentiated analysis seems to be necessary to find out which position market prices have in relation to corporate values. For this purpose the further analysis starts with the assumption of a perfect capital market and then gradually loosens the strict assumptions until more realistic settings are reached.

A perfect market should be defined as a market without frictions (no transaction costs, no arbitrage costs, no taxes, assets are perfectly divisible and marketable, no restrictions on shortselling), with perfect competition (all investors are virtual price takers[15]), with informational efficiency and utility maximizing individuals (see Copeland et al., 2005: 353-354; Gerke and Bank, 2003: 61; Hirshleifer and Hirshleifer, 1997: 410-411; Hirshleifer, 1958: 330; Fama and Miller, 1972: 21-22; Dothan, 1990: 20). Additionally, assuming that non-financial

---

[14] It should be abstracted from the situation in which substantial stakes of companies are traded, e.g. in M&A-transactions or initial public offerings (IPOs).

[15] In general, in perfect markets all investors are factual price takers. However, based on this assumption no reasonable conclusions can be drawn concerning the relationship between value and price. That is why this strict assumption is abandoned here.

goals do not influence the valuation process of individuals, the differences in their expectations should be rather small. Thus, in such a market the resulting price would always equal the (subjective) decision value of a hypothetical, typical average-investor (see Pratt et al., 2000: 30-31; Prokop, (2003: 49). Consequently, in a (virtually) perfect market the majority of investor's decision values can be expected to be close to the market price. Since each investor typically has a certain decision interval – i.e. a range of decision values (see Hering, 1999: 91) – the price determined in a (virtually) perfect capital market can serve as a decision value for many of these investors. This price can also be seen as an objectified value following the Anglo-Saxon interpretation (see Pratt et al., 2000: 28-30). Contrary to this, the price in a perfect market is not an objectified value in the sense of the IDW. The IDW interpretation is that the objectified value equates a seller's value (see IDW, 2000: 831-832), while this price incorporates both sell-side and buy-side expectations. It should be noted that this price does not equal the "objective value" either, since investors are only virtual price takers, and in fact have their own expectations which become manifest in their marginal prices.

Loosening some of the strict assumptions and approaching the real capital market settings gives rise to a weakening conformity between value and price. While a high degree of divisibility and marketability is given even in real capital markets (see Campell et al., 1997: 9; Prokop, 2003: 52), the existence of transaction costs and taxes can no longer be suppressed. However, transaction costs are not a differentiating factor since they are part of every transaction. Additionally, if in a special valuation case transaction costs are assumed to deviate from the typical capital market transaction cost, an adjustment is easily possible. The provision for taxes also does not dramatically change the relationship between value and price. In Germany and most other developed nations companies have to pay corporate taxes irrespective of investor's characteristics. Thus, only personal income taxes might possibly make a difference between value and price. In this context it could be shown that there must be identity between pre- and after-tax valuation as long as investors believe in neoclassical asset pricing models such as the standard Capital Asset Pricing Model (CAPM)[16] or the after-tax CAPM (see Peemöller et al., 2005). Minor differences between both value dimensions, however, might arise if investors are forced to leave the CAPM-world (e.g. when valuing private firms) (see Copeland et al., 2000: 153; IDW, 2002: 42; Schultze, 2003: 286-289, 312; a different opinion is provided by Damodaran, 1996: 112). The latter point is, however, not of major importance here because the discussion is clearly about capital market oriented valuation. Anyway, the Anglo-Saxon valuation practice widely waives the provision for personal income taxes for practical reasons (see IDW, 1998: 68; Schultze, 2003: 286). This leads to the supposition that – if the respective investor does not face a situation of extremely high or extremely low taxes or the investor leaves the CAPM world – the impact of personal income taxes on decision values is not significant.

---

16 For more information about the CAPM, see Footnote 25. For a discussion about what „believing in the CAPM" means on real capital markets considering psychological issues, see Gerke (1997).

Of course, the degree of market efficiency is crucial for the explanatory power of market prices. At this point it should suffice to assume that on the German capital market information is more or less mirrored in market prices and that there are in general no major distortions due to informational inefficiencies.[17] It also has to be considered that especially if investors are not perfectly informed, the market price can sometimes serve as a better decision value than the marginal price that investors calculate based on their own forecasts. This is possible since investors' partial information might be aggregated in market prices and therefore the market's estimates about the future development of a company or an industry are superior to the individual's forecasts.[18] However, for stocks that obviously lack liquidity and for which trading occurs infrequently, it is questionable whether the market price has any significance (see Hommel and Braun, 2002).

Finally, one feature with a very strong impact on the relationship between price and value is that investors' expectations can no longer be regarded as quasi-homogeneous. Not necessarily every investor on real capital markets values a company for investment reasons. For example, there might be some market participants that act as strategic buyers[19] – and therefore include possible synergies into their valuations – or at least those which hope that strategic buyers enter the market (noise trader) (see Black, 1986; Kyle, 1989; De Long et al., 1990b; Shleifer and Summers, 1990: 19-33; Menkhoff and Roeckemann, 1994: 277-295. A different view on noise traders is provided by Gerke, 1997). Also, certain "myopic" institutional investors, whose investment horizon is rather short, might undervalue the importance of distant cash flows of the target company.[20] This situation of varying investors' expectations generally leads to a wide variation of the decision values of market participants. Attention should also be turned to the fact that market prices are usually minority prices, i.e. they are not the relevant prices for investors that want to buy a bigger stake of a company to exercise control (see Hering, 1999: 94-95; Prokop, 2003: 50; Bamberger, 1999: 667-668). However, in many cases it is possible to adjust market prices to the interests of those controlling investors (see Pratt, 2001: 136-144. Regarding control premiums, see Shleifer and Vishny, 1986: 461-488; Franks et al., 1988: 234, 242; Boecking and Nowak, 1999: 173-174; see also section 2.4.3).

---

17 See section 3.1.2 for an in-depth view on capital market efficiency.

18 See Rubinstein (1974: 225-244); Gerke and Rapp (1994: 11-12); Pesendorfer and Swinkels (2000: 499-525); see Weber and Wüstemann (2004: 6-8) for a practitioners' support for this thesis; see also section 3.1.2 for more information about aggregation efficiency of stock markets.

19 Strategic buyers usually expect synergies from acquisitions while financial buyers consider acquisitions as investments; regarding the deviant marginal prices of such strategic buyers, see Bhagat and Hirshleifer (1996), Hietala et al. (2000).

20 See Porter (1992: 65-82); Lang and McNichols (1997); Abarbanell and Bernard (2001: 221-242); Bushee (2001: 207-246). In some cases even management behaviour seems to be myopic, see McConnell and Wahal (1997). Studies that rather support the irrelevance of the myopia-thesis are provided by McConnell and Muscarella (1985: 399-422) and Wooldridge and Snow (1990: 353-363).

From the discussions above it can be seen that, despite the weaker conformity, the identity of market prices and *decision values* (on average) is still possible assuming a more realistic environment with several market imperfections (see Prokop, 2003: 64; Ruhnke, 2003: 92; Großfeld, 2000: 261-262; Böcking, 2003: 84-85; a contrarian view is provided by Hering, 2000: 441-443). More precisely, the market price still equals the decision value of the typical average (minority) investor and therefore should be valid for many market participants.[21] However, investors have to thoroughly analyse the dimension of each imperfection and its respective impact on the explanatory power of market prices. It is even more important to note that for investors with intentions totally different from the average market opinion – such as a strategic buyer who expects synergy effects from the purchase of a whole company while no takeover scenarios are currently priced – neither the unadjusted nor an "across the board"-adjusted price can serve as a decision value (see Hering, 2000: 441). Nevertheless, even if the expectations of the appraiser and those inherent in the market price differ, the market price can serve at least as a plausibility check for individual decision values (see Herrmann, 2002: 16; Pratt et al., 2000: 30-31; Bausch, 2000: 457).

The usability of market prices as a *neutral-referee value* must also be seen in light of the factual degree of market perfectness. While in general the IDW stipulates that market prices must not remain unconsidered in such valuation settings, it also specifies that any factors that could give rise to distortions of prices – such as lack of marketability or the obvious existence of speculative attacks – have to be carefully examined; in the case of existing distortions this definitely leads to the uselessness of market prices for this purpose (see IDW, 2000: 828-829).

Independently of the state of the capital market, the market price can always be seen as an *argument value* (see Hering, 1999: 177; Prokop, 2003: 64). This is due to the growing importance of market prices in business – in terms of the shareholder value perspective on management or the emergence of stock news in the press – but also to the high reputation of certain academic capital market research (see Hering, 1999: 188). As a matter of course, when the market price is within the interval of the lower limit of the seller's prices up to the upper limit of the buyer's prices, it can serve as an *intermediation value* as well (see Hering, 2000: 450; Schildbach, 1998: 319).

In conclusion: The usefulness of "market price" as a synonym for "corporate value" crucially depends on the function that the corporate value has to fulfil and on how developed the capital market is. Since the difference between the observable market price and the (unobservable) corporate value is not measurable (see Prokop, 2003: 54-55), the decision about this usefulness has to rely on the subjective assessment of a skilled appraiser. Regardless of this, the usefulness of market prices as part of the valuation process is great anyway because of their appropri-

---

21 This is consistent with the view of the US-GAAP standard setter, the Financial Accounting Standards Board (FASB): „An observed market price encompasses the consensus view of all marketplace participants about an asset or a liability's utility, future cash flows, the uncertainties surrounding those cash flows, and the amount that marketplace participants demand for bearing those uncertainties" (FASB, 2002: 304).

ateness as a plausibility check and as a means of balancing the interests of minority shareholders (e.g. in a "squeeze-out") as well as, of course, because of the need for plurality of valuation approaches.

2.2.3 Approaches to Company Valuation

Valuation theory knows three general approaches to estimate the value of companies: direct valuation, contingent claim valuation and relative valuation (see Bhojraj and Lee, 2002: 413-414; Damodaran, 2002: 11). A fourth approach – called the asset based or substance based approach – is sometimes used under special circumstances. Each approach requires a different set of input variables, a different forecasting scope and follows different assumptions. At first glance it even seems as if these approaches do not have much in common, but in fact they share many similarities. It is necessary to have a good grip of the basics of the different valuation models and techniques to understand how CCV fits into the big picture. Therefore, these four approaches are described briefly in the following section.

2.2.3.1 Direct Valuation

This approach (also sometimes called the *income approach*) directly determines the corporate value on the basis of what investors can get out of the company in the future. It is based on the net present value (NPV) concept, which basically states that in a world of uncertainty the value of a project, respectively a company, equals the sum of all expected future risky financial benefits[22] discounted at the opportunity cost of capital:[23]

$$V_t = \lim_{N \to \infty} \sum_{n=1}^{N} E\left[FB_{t+n}\right] \cdot \left(1 + c_{t+n}\right)^{-n} \tag{2.1}$$

---

22 Financial benefits are sometimes also called "economic benefits", see ASA (2002: 24); however, the term "financial benefits" is used throughout this study in order to clarify that the focus here is on financial inflows.

23 See Copeland et al. (2005: 881-883); Kothari (2001: 4); Brealy and Myers (2003: 14-18); Damodaran (2002: 11-12); Gerke and Bank (2003: 42-43); this concept is also known as the dividend discount model. Another method of directly determining the corporate value is to apply the certainty equivalent approach, where – given an investor's utility function – the (downward adjusted) risk free amount of financial benefits, that yields the same utility to an investor as the (higher) risky amount, is discounted at the risk free rate, see IDW (2000: 833); Peemöller and Kunowski (2005: 235-236); Mandl and Rabel (1997: 218-225). However, the latter concept is not subject of the following discussions.

with: V_t = Value of the company as a whole at time t
FB_{t+n} = financial benefits to the investors at time $t+n$
c_{t+n} = cost of capital at time $t+n$

Assuming that current financial benefits grow at a constant rate and that the cost of capital remains constant over time, Equation 2.1 can be simplified to the so-called "Gordon growth model", where corporate value is defined as the next year's expected amount of financial benefit divided by the difference between the discount rate and the long-term growth rate of financial benefits (see Williams, 1938; Gordon and Shapiro, 1956: 102-110; Brealy and Myers, 2003: 64-65).

$$V_t = \left(FB_t\right)\cdot\left(1+g\right)\cdot\left(c_c - g\right)^{-1} = E\left[FB_{t+1}\right]\cdot\left(c-g\right)^{-1} \quad (2.2)$$

with g = sustainable growth rate of financial benefits

This very compact model also illustrates the connection between the corporate value and current fundamentals of that company. It is worth noting that the only forecast necessary when using the Gordon growth model is that of a plausible growth rate. Usually, in valuation practice a detailed forecast of financial benefits is performed for 3 to 5 years, while a simplified forecast using the Gordon growth model is performed for the remaining years (see Copeland et al., 2000: 234; IDW, 2002: 61).

Since the practical use of direct valuation is widespread but of many variants, some light should be shed on the two main categories of this approach: the internationally accepted discounted cash flow (DCF) methods and the German "Ertragswert" method. Additionally, a short description of the residual income concept will be provided.

All of the DCF methods have in common that the numerator comprises risky cash flows. For further analysis, two DCF valuation variants are of particular interest: the entity approach and the equity approach.[24] In the entity approach – also known as the weighted average cost of capital (WACC) approach – expected free cash flows to the firm are discounted. Free cash flows to the firm are the cash flows available to all capital providers (debt and equity) after net investment but before financing activities (see Copeland et al., 2000: 167-171; Brealy and Myers, 2003: 74-75; for companies that are taxable in Germany see Baetge et al., 2005: 283-284; Hachmeister, 2000: 61-72). These cash flows are the financial benefits that matter to investors. The discount rate applied in the entity approach equals the average cost of equity and debt capital weighted by the respective market values. The cost of equity is usually determined using the Capital Asset Pricing Model (CAPM) (see Kuelpmann, 2004: 4-5), a model used to value risky assets in case of

24 For a detailed description of the DCF-concept with its various variants see Copeland et al. (2000: 129-297); Drukarczyk (2003: 199-314); Mandl and Rabel (1997: 37-42); Baetge et al. (2005: 265-362).

capital market equilibrium.[25] Finally, to determine the value of ownership of the company (equity value), the market value of debt has to be deducted from the (enterprise) value.

Contrary to this indirect calculation of the shareholder value, the equity approach directly focuses on cash flows available to shareholders. Thus, financing activities are explicitly considered when determining the cash flows (see Copeland et al., 2000: 150-152, concretely for companies that are taxable in Germany see Baetge et al., 2005: 276-277, 285-286). To determine the equity value, the free cash flows to the owners have to be discounted at the cost of equity, which (again) is usually calculated using the CAPM.

The "Ertragswert" method (following the standards of IDW) is a form of earnings capitalisation model, which is especially popular in Germany (see IDW, 2000: 835-837; Peemöller and Kunowski, 2005: 201-263). This valuation model discounts future earnings at the cost of equity to yield the equity value.[26] Even if earnings are rather an accounting figure than a financial benefit, earnings capitalisation models in general can be conclusively transformed from Equation 2.1 (the dividend discount model) (see Fama and Miller, 1972: 87-89; Campbell and Shiller, 1988a: 661-676; Campbell and Shiller, 1988b: 195-227; Fama, 1996: 415-428). This transformation shows that a major problem of earnings capitalisation models is that earnings not only reflect the value added due to profitable future investments but also growth due to the reinvestment of earnings (see Fama and Miller, 1972: 88-89; Kothari, 2001: 75). The "Ertragswert" method tries to overcome this drawback by assuming that earnings are fully distributed to shareholders on the condition that the asset base and the capital structure of the company are maintained.[27] Thus, from a purely technical point of view, the "Ertragswert" method is similar to the equity approach of the DCF model (see Drukarczyk, 2003: 304-305). Differences between both approaches exist in the assumption

---

[25] The CAPM can be traced back to Sharpe (1964), Lintner (1965) and Mossin (1966). See also Gerke (2001, colums 1701-1703); Copeland et al. (2005: 147-176); Brealy and Myers (2003: 194-204); Gerke and Bank (2003: 242-250); for a discussion about the shortcomings and limitations of pracital applicability in the context of business valuation see e.g. Hering (2003: 289-296); Fama and French (2003). In Germany, the application of a modified CAPM in direct valuation approaches is currently being discussed, which is supposed to better account for the impact of personal income taxes when determining objectified corporate values (after tax-CAPM); see Brennan (1970); Wiese (2004); Jonas et al. (2004); Wagner et al. (2004); IDW (2005). A critical evaluation of this proposition can be found in Peemöller et al. (2005). In the USA the cost of equity is sometimes determined using a multi-factor approach in the sense of Fama and French (1992) and Fama and French (1993), see Stehle (2004: 914), or using the arbitrage pricing theory (Ross, 1976). Other approaches discussed in literature – such as the intertemporal CAPM (Merton, 1973) or the Zero-Beta CAPM (Black, 1972) – are of no major practical relevance.

[26] The cost of equity can be, but does not necessarily have to be, determined using the CAPM, see IDW (2000: 834, 836-837).

[27] See Peemöller and Kunowski (2005: 217); IDW (2000: 830); consequently all future profitable investments are assumed to be debt-financed.

about how future projects are financed, and – potentially – in the determination of the cost of equity (see Eidel, 2000: 43-44; IDW, 2002: 836-838).

Residual income valuation (RIV) is also based on the NPV concept. However, this approach additionally assumes that the book value of equity serves as a proxy for the present value of all future normal earnings. In this context, normal earnings are defined as those earnings that yield a zero economic profit. The equity value VE is then determined by summing up the book value of equity and the present value of future abnormal earnings. These future abnormal earnings are calculated as the difference between forecasted earnings and normal earnings[28]:

$$VE_t = EQUT_t + \lim_{N \to \infty} \sum_{n=1}^{N} \left(NI_{t+n} - c_e \cdot EQUT_{t+n-1} \right) \cdot \left(1 + c_e \right)^{-(t+n)} \qquad (2.3)$$

with $EQUT_t$ = accounting book value of equity at time t
NI_t = net income at time t
c_e = cost of equity

Equation 2.3 can be directly transformed from Equation 2.1 (see Fernandez, 2003). While the RIV concept has been around for a long time (see Preinreich, 1938: 240; Solomons, 1965: 126-127), it has become popular only in recent years. Especially, the Economic Value Added (EVA™) model and the Ohlson model respectively the Feltham-Ohlson model have successfully renewed this valuation idea. The EVA™ concept can be regarded as some kind of adjusted RIV model that explicitly aims to determine market values (see Sheehan, 1994; O'Byrne 1997; Hostettler, 2000: 19-37). A more detailed description of the Ohlson model can be found in section 3.2.1.3.

A common part of all direct valuation approaches is that a thorough forecasting ability is required to perform business valuation accurately.[29] In general, the prediction of future cash flows is regarded as more difficult than forecasting earnings (see Penman and Sougiannis, 1998). Additionally, comparing RIV models with conventional earnings capitalisation models reveals that in the case of RIV bad predictions lead to lower variability of the calculated corporate values because of the smaller impact of future earnings on the terminal value. Usually, the lesser weight a model places on the terminal value in the calculation of value, the more it is preferred in the sense of valuation accuracy (see Penman, 1998).

---

28 See Lo and Lys (2000a); Schumann (2005: 22-23); the general opinion is that for the validity of this model, clean surplus accounting is required, see Edward and Bell (1961, 68); Peasnell (1982, 362). The clean surplus relation means that only capital contributions, dividends and the profit or loss reported in the income statement can change the amount of owners' equity (see Footnote 58 for some more details about the clean surplus relation). Recently, however, it has been found that residual income valuation is still a reasonable valuation approach even if the accounting system is not perfectly clean surplus, see Yee (2005).

29 Lee (1999: 414) states in this context that the "essential task in valuation is forecasting. It is the forecast that breathes life into a valuation model". See also Kothari (2001: 72-73).

2.2.3.2 Contingent Claim Valuation

Contingent claim valuation describes a valuation approach for certain assets for which cash flows are not fixed but depend on the occurrence or non-occurence of an event. In the simplest case this event is the arrival of new information which might influence the decision whether or not to invest in this asset. Thus, the managers of a company have some kind of future flexibility on how to respond to this new information (see Lander and Pinches, 1998: 537-538; Myers, 1977: 155; Myers and Majd, 1990: 3; Amram and Kulatilaka, 1999: 96-97; McDonald and Siegel, 2001: 253, 255-256). Common direct valuation models usually do not capture this flexibility appropriately. The determined present value is the expected value of a symmetrical normal probability distribution – at least when using the CAPM for determining the discount rate (a different opinion is provided by Fama, 1996: 416). Adequately accounting for flexibility, however, requires more weight to be placed on the upside potential of the project value while attaching less importance to the downside losses. Doing this leads to a higher expected value and causes the probability distribution to be skewed to the right (see Trigeorgis, 1988: 147; Trigeorgis, 1996: 123).

This positively skewed distribution of the project value strongly resembles the typical distributions of financial options. In fact, contingent claim valuation is a reasonable approach for assets that exhibit optionlike features. Due to this conceptional analogy it requires the use of variants of financial option pricing models such as the Black/Scholes/Merton (BSM) model or the binomial model.[30] For this reason these assets/projects are called real options and contingent claim valuation is also known under the name of real options valuation.

It is worth noting that, along with flexibility, the existence of a time period in which management can benefit from this flexibility is also a major characteristic of real options (i.e. the option must have time value) (see Peemöller and Beckmann, 2005: 805). Classical examples of optionlike assets are patents, real estate, research and development projects, undeveloped natural resource reserves and human resources (see Damodaran, 2002: 22-23; Lander and Pinches, 1998: 539). Moreover, any asset or project can be valued using the real option approach if management has the flexibility to defer, abandon or otherwise alter that project in the course of time (see Trigeorgis, 1996: 9-14; Copeland et al., 2000: 400-402). The limit of the practical applicability of this approach lies in the similarity to financial options and in the data availability (see Peemöller and Beckmann, 2005: 807). Forecasting problems especially arise in determining the volatility of the project's cash flows.

---

30 For option pricing models in general see Hull (2003: 234-329); in most cases the basic models for financial options have to be adjusted to the specific real option; see Quigg (1993), who developed a model for valuing real estate assets by modifying the BSM model in that the exercise price is stochastic; see also Gibson and Schwartz (1990), who generated a model for the valuation of commodity related projects. They substituted a mean reverting process for the standard Brownian motion in the BSM model; see also Schwartz (1997) and Bjerksund (1991), who provided an analytical solution for this model.

2.2.3.3 Relative Valuation

The relative valuation approaches include the use of "rule of thumb" multiples based on personal experience and the use of multiples derived from comparable (mostly publicly listed) companies.[31] While the first variant is a very subjective and simplified approach and without any deeper theoretical foundation, the latter plays a major role in modern valuation settings. The idea of CCV arises from the basic principle of arbitrage, which states that all substitutes should sell for the same price (see Gerke and Bank, 1998: 231-232). In the context of business valuation this means that – given uniform investors' expectations and valuation circumstances – equal companies should have the same value. Usually no explicit direct valuation is performed for the comparable company but rather market prices are used as reference. Thus, the objective of this approach is to value target companies based on how the market prices similar companies. Usually, the market prices of comparable companies are derived from one of the three following categories: recent acquisitions, recent initial public offerings or stock listings of public companies (see Mandl and Rabel, 1997: 259-265; Löhnert and Böckmann, 2005: 405). The latter variant is called the market approach to CCV.

In the simplest case, perfect substitutes exist for the target company. In such a situation the valuation is straightforward (this approach is called immediate CCV). In real capital markets, however, companies are typically not identical. As a consequence, relative valuations have to rely mostly on similar companies whose market prices have to be "adjusted" to yield the value of the target company. This adjustment is done by considering the relation of certain financial or non-financial key figures, in which the two companies differ. Single-factor and multi-factor comparable company valuation is typically distinguished, dependent on the amount of key variables included in the valuation model. Single-factor models are by far the dominating CCV approach in practice (see Fernandez, 2002; for Germany see Kames, 2000: 58-60, 100-101; Wichels, 2002: 146, 148; Duerr, 1995: 27; Kusterer, 2003: 99-100), while the practical relevance of multi-factor models is rather limited thus far.[32] To account for minor mispricing or differences in the similarity, a set of companies (the so-called peer group), whose adjusted market values are aggregated to yield an average adjusted market value, is used as a standard of comparison in most real valuation settings.

From a methodological point of view, following the CCV approach, the value of a company is a function of a certain number of bases of reference weighted with the respective sensitivity factors. Each sensitivity factor is drawn based on the relationship between the basis of reference and the market price of the comparable companies. The basis of reference is determined either at the time of the

---

31 See Mandl and Rabel (1997: 42-46). "Rule of thumb" multiples for several German industries can be found at www.finance-magazin.de. It is important to note that value indications derived from these "rule of thumb" multiples are probably not very meaningful, see ASA (2002: 13).

32 See section 4.3 for an overview of the practical relevance of multi-factor models.

valuation *t* (trailing CCV) or is the expectation value of the key figure at a later point in time (leading CCV):

$$V_t = \sum_{n=1}^{x} a_{nt} \cdot BR_{n\tau} \quad (2.4)$$

with τ = point in time that equals *t* or later, i.e. $\tau \geq t$

$BR_{n\tau}$ = n^{th} basis of reference (financial or non-financial key figure), determined at time τ.

a_{nt} = sensitivity factor for $BR_{n,\tau}$, determined at time t.

x = number of bases of reference applied; if x = 1 then the approach is single-factor CCV, if x ≥ 1 then the approach is multi-factor CCV.

Since this approach relies on market prices, no *explicit* forecasts of future cash flows, earnings or discount rates are required. However, this does not mean that future prospects are not relevant for comparable company valuation. In fact, forecasting problems are shifted from explicit estimates of single companies' future business development (in the case of direct valuation) to the prediction of future similarity between the target company and the comparable companies. It remains to be seen whether this shift effectively leads to a reduction of the high demands on the forecasting ability of appraisers.

2.2.3.4 Asset-Based Valuation

A special form of direct valuation is the asset-based approach (also called the substance-based approach). It is a general way of determining a value indication based on the value of a company's assets net of liabilities (see ASA, 2002: 8). This approach does not really fit into the valuation framework provided by the other three valuation approaches. The asset based approach is not a general valuation method. In fact, it is rather appropriate in case a company does not continue its current business activities. In the simplest case, a company plans to change current business activities by adapting its resources to different uses (see Berger et al., 1996; Burgstahler and Dichev, 1997; Collins et al., 1999; Sieben and Maltry, 2005: 380-385). The variant of asset-based valuation that has to be applied then is called reproduction valuation (also called reconstruction valuation, net-asset valuation or cost approach to business valuation, see Barthel, 2002: 201-202; ASA, 2002: 23). It requires all assets to be valued at replacement costs.[33] The most common case for the application of asset-based valuation is when a company faces

33 Reproduction valuation itself can be divided into different variants. It is also sometimes used for other purposes, such as valuation of non-listed corporations, valuation for tax purposes (in Germany, e.g. as part of a combination model of reproduction valuation and earnings capitalisation – the so-called "Stuttgarter Verfahren") or in certain disputes. From an economic perspective, however, this approach is subject to critique, see Sieben and Maltry (2005: 400).

bankruptcy and has to be liquidated. This part of asset-based valuation is called liquidation valuation (see Sieben and Maltry, 2005: 398-399; Schultze, 2003: 152; ASA, 2002: 26). Under this variant, single assets are valued based on what the owner of the company could expect if he is forced to sell these assets in liquidation (see IDW, 2002: 127-128; Barthel, 2002: 204-205; as regards the costs of liquidation or bankruptcy, see Richter, 2002: 307-311). A third variant, which has found some support in Germany, understands the assets-in-place as some sort of expenses already paid. This approach is not a tool to determine the value of a company but rather delivers additional information for other valuation approaches (see Sieben and Maltry, 2005: 398). It is worth noting that the liquidation valuation and the assets-as-expenses-already-paid approach require certain – even if limited – forecasting ability (see Sieben and Maltry, 2005: 400). Last but not least, shareholders' equity (directly drawn from the financial statement) can also be understood as an asset-based value (see Schultze, 2003: 152; Barthel, 2002: 202).

2.2.4 Purposes of Appraisal

There is a myriad of reasons for performing business valuations. Along with those required by law, there are multiple motivations to value a company depending on the life cycle stage of that company and on the intention of (potential) investors. Additionally, all purposes can be partitioned into those that are associated with a (planned) change of ownership and those where ownership structure remains untouched. In the latter context it is important to note that, even if a factual transaction does not take place or is not planned, the perspective of the person who values the company can always be seen as either a sell-side position or a buy-side position.

Figure 1 abstracts and shows graphically the most important reasons for performing valuations classified as the categories mentioned above. Some reasons appear in the table more than once because they can be classified into more than one category. Nevertheless, this table is not exhaustive and it is emphasized that it only contains some of the most common examples. The classification according to the type of regulation exclusively focuses on legal requirements especially relevant to companies operating in Germany.

Fig. 1. Classification scheme of purposes of appraisal

According to company's life cycle	According to the type of regulation	According to the intention of (potential) investors	According to the ownership of the company
• Formation of a company • Initial Public Offering (IPO) • Going Private • Mergers & Acquisitions • Divestitures • Restructuring • Liquidation of the company	Regulation by Law *German Stock Corporations Act (AktG):* - Prerequisite for the conclusion of certain agreements (profit transfer agreement, domination agreement) - Integration of a company into another corporation - Squeeze-out, minority oppression action *German Transformation Act (UmwG):* - Transfer of assets and liabilities of a whole entity to another entity (Verschmelzung) - Split-ups, split-offs, spin-offs *Other legal regulations:* - Impairment test - Tax purposes Regulation by Contract - Entry and exit of partners of a partnership - Heritage disputes - Martial dissolutions Internal regulations of a company - e.g. value-based management which includes periodical performance controls	• Purchase and sale of whole companies • Purchase and sale of interests in a company • Entry or exit of partners of a partnership • Contribution of assets or business units to a company	Change of ownership - Dominated situations: One party can enforce its claim[a] - Un-dominated situations: Mutual consent of both parties is necessary[a] No change of ownership - Value-based management - Recovery, restructuring - Credit rating - Inheritance tax

[a] see Matschke, 1976: 26-39 for an in-depth discussion of both situations

Figure according to Peemöller (2005b: 17-21), Pratt, Reilly and Schweis (2000: 34).

2.2.5 Classification of Comparable Company Valuation

Interpretation of valuation results

Obviously, CCV is predominantly a relative valuation approach to determine market prices of companies. However, the term "market price" is not totally independent from, but rather connected to the term "corporate value". As already discussed, market price can by all means be seen as identical to certain interpretations of corporate value or can at least be transformed to any form of corporate value. Therefore, market prices must be defined as a subset of corporate values and consequently price determination is not the only scope of CCV. The concrete aptitude of CCV for business valuation purposes is closely related to the degree by which market prices can be seen as a synonym for corporate values.

Even if sometimes misleadingly stated (see e.g. Olbrich, 2000: 458), CCV does not aim to determine an objective value of a company, since this approach is based on market estimates and therefore not independent of investors' preferences. On the contrary, the outcome of the valuation process is a subjective value from the perspective of an "average" market participant with "average" expectations and expectations about the future development of the target company.[34] Concretely, the result is an objectified value – not in the reference to the IDW-interpretation, which rather means a sellers' value, but as defined in the Anglo-Saxon valuation theory: a value from the perspective of a typical average investor.

From the standpoint of the theory of functional value, CCV can be a tool to determine argument values for any party involved in the valuation process. On the condition that the valuation is accurately and properly conducted, CCV represents a valid means to underpin the reasoning of market participants.

In developed capital markets, CCV might also provide decision values for market participants. However, a detailed examination of the factors that drive the market price of the comparable companies and the target companies is necessary: Only if the appraiser's expectations do not deviate too much from the expectations implied by the market price, CCV is a powerful tool to determine decision values. This is especially important in case of a pending M&A-transaction, where the buy-side objectives on how to proceed with the target company after the deal are frequently quite different from the current market expectations. It is important to notice that, in the case where there are only small deviations between market expectations and appraiser's expectations, CCV can still provide decision values if the appraiser feels that the advantage from immediate value determination at low informational costs more than offsets the disadvantage of minor valuation inaccuracies. Additionally, certain discrepancies between market prices and decision values, such as the lack of a control premium in market prices, can largely be balanced using experience values. Nevertheless, if market prices are derived from recent acquisitions or recent IPOs, the use of CCV to determine decision prices is very limited due to obvious market imperfections.

[34] See the interpretation of market prices from the FASB in Footnote 21 (FASB, 2002: 304).

If, finally, the outcome of the CCV process lies above the lower limit of the seller's decision value and below the upper limit of the buyer's decision value, it can serve as a neutral referee value in litigations.

Link to other valuation approaches

Direct valuation methods, especially DCF-approaches, are a means of estimating market prices (see FASB, 2002: 301; Herrmann, 2002: 26-28; Drukarczyk, 2003: 130). By determining the discount rate based on market equilibrium models, such as the CAPM, DCF-approaches try to imitate the circumstances at real capital markets. From a methodological point of view, the intention of these models is to calculate corporate values as seen through the eyes of a typical, capital market-oriented "average" investor (see FASB, 2002: 304; Prokop, 2003: 50-51). Certainly, DCF-approaches allow for individual subjectivity when determining the cash flows and therefore are not inevitably bound to determine market prices.[35] However, strictly speaking, in this case it is not the value of the target to a special investor that is calculated, rather a potential price which the capital market (i.e. the average investor) would assign to the target if the respective cash flow scenario were true.[36]

Anyway, the capital market aggregates these different cash flow scenarios to form the factual market price and, thus, market prices can be seen as a function of investors' direct valuation approaches, especially of DCF-approaches. As a consequence, direct valuation approaches are implicitly the basis of CCV and therefore inseperably connected to CCV.[37] For classical single-factor models, such as the price-earnings (PE) ratio, this relationship can technically be expressed via the Gordon/Shapiro equation, which is derived from the Gordon growth model (Equation 2.2) (see Beaver and Morse, 1978: 65; Damodaran, 2002: 471):

---

35 See Casey (2000: 19-20). The degree of subjectivity might even be higher for earnings capitalisation methods like the IDW-"Ertragswert" approach since it allows the discount rate to be based on investor's preferences and degree of risk aversion, see IDW (2000: 837). That is why they are usually not seen as means of estimating market prices; see Drukarczyk (2003: 130); Herrmann (2002: 33). However, under certain conditions the use of the CAPM is also acceptable to determine discount rates, see IDW (2000: 836-837 in conjunction with p. 834).

36 The determination of a value from the perspective of an individual investor would require using a discount rate that mirrors the individual risk aversion and investment situation (degree of diversification of the portfolio, planned holding period, etc.), see Hering (2003: 289-292, 343-344).

37 It is important to note that direct valuation approaches and CCV are similar in their methodology, too. Direct valuation approaches also require the existence of a (at least virtual) comparable investment to determine the discount rate, see Gerke and Bank (2003: 100-101); Schultze (2003: 63-70); e.g., if the CAPM is used for determining the cost of equity, the comparable investment consists of two securities, the riskless asset and the beta weighted market portfolio. As a consequence of this methodological similarity, literature sometimes calls the problems of investment comparability that are associated with direct valuation approaches at least as severe as those associated with CCV, see Richter (2003: 312).

$$PRICE_t \cdot \left(NI_{t+1}\right)^{-1} = \left(DIV_{t+1} / NI_{t+1}\right) \cdot \left(c_e - g\right)^{-1} \qquad (2.5)$$

with $PRICE_t$ = market price of the company at time t
DIV_{t+1} = dividends of the company for the period t+1

From this equation, it can be seen that certain multiples in CCV approaches (here: the PE ratio on the left hand side of the equation) can be directly explained by some form of adjusted direct valuation approach (on the right hand side of the equation: the Gordon growth model with dividends scaled by earnings).[38] More generally, the relationship between CCV and direct valuation approaches can be expressed as:

$$V_t = \sum_{n=1}^{x} a_{nt} \cdot BR_{n\tau} = f\left(VF_{direct\ valuation}\right) \qquad (2.6)$$

with $VF_{direct\ valuation}$ = the value determining factors for direct valuation approaches

Since contingent claim valuation is often performed along with direct valuation, the value determining factors for direct valuations include potential real options of the target company. Moreover, contingent claim valuation must also be seen as a variant of relative valuation because it is also based on already existing (potential) market prices. Thus, all three fundamental valuation approaches are closely connected. However, while direct valuation and contingent claim valuation can exist and reasonably be performed without CCV, the opposite is not true. CCV crucially relies on the existence of direct valuations (including real options) already performed by market participants, and a successfully functioning market that accurately aggregates the valuation results of single investors.

Application range

The interpretation of market prices as potential decision values, arbitrium values, and neutral referee values also enlarges the application range of CCV from pure price determination to more complex valuation tasks. Certainly, the primary objective of CCV remains to assign prices to companies that do not have a quoted market price yet, e.g. in the case of IPOs and in part in the case of M&A-transactions (see Nelles et al., 2001: 323-324; Bausch, 2000: 456-457). Furthermore, CCV is a tool to verify if the equity of listed companies is over- or undervalued, e.g. in the case of stock valuation reports. In this context, CCV can also be used to determine the fair market value of such equities (see e.g. Wichels, 2002: 146, 148). Additionally, it can be applied in many other valuation settings described in section 2.2.4 and can fulfil functions described at the top of this section, as long as the al-

38 See section 3.1.1 for an in-depth presentation of how other single-factor CCV models are related to direct valuation models.

ready mentioned requirements concerning the market prices of the comparable companies are met.

Nevertheless, independent of the quality of the market prices, there are certain valuation settings that categorically do not favour the application of CCV. On the one hand there are legal restrictions: In Germany, the largest part of valuation causes induced by law requires the exclusive use or at least the principal use of direct valuation approaches (see Piltz, 2005: 783-788. In the USA, however, CCV has some importance in legal practice). On the other hand there are economic restrictions: CCV has been proven to be rather useless when the impact of certain strategies on market prices should be assessed, as in the case of value based management. Additionally, most strategic buyers do not profit from CCV when determining the decision value because of their need for taking exclusive strategy changes and potential synergies into account (see Bausch, 2000: 454-456; Hering, 2000: 441). However, CCV can be used to supplement other (direct) valuation approaches of strategic buyers. It might help to see how the market would value a company with no specific strategy change implemented. Then the potential value-added from a planned new strategy can be determined as the difference between the strategic value and the current stand-alone value. After all, the supplementary application of CCV may be appropriate in many cases, just because of the need for plurality in company valuation methods.

2.3 Rationale and Style of Comparable Company Valuation

2.3.1 Immediate Valuation Models

The various shapes that CCV may take on can all be traced back to the basic valuation model, the so-called *immediate valuation model*. Applying this model means directly projecting the value of one company (or the average value of a group of companies) to the value of the target company without the use of any reference variables. Although this model is of almost no use at all in practice for reasons to be outlined below, it is important to give a brief explanation in order to provide a better understanding of the fundamental idea, the proceeding and the aim of CCV in general and to deliver a basis for more complex approaches of CCV.

The immediate valuation model goes back to one of the basic principles of financial theory in general and valuation theory in particular: the principle of arbitrage. This principle simply says that all substitutes should sell for the same price. If that is (temporarily) not the case, investors have the opportunity to buy the cheap asset while correspondingly selling the rich asset and thus earning a riskless profit (see Gerke and Bank, 2003: 270-271). This trading strategy – called arbitrage – will force the price spread to narrow and finally lead again to price equality.

In the context of business valuation this means that equal companies should have the same value.[39] Equality of companies especially means that all value-driving factors have to be identical. Thus, taking the NPV concept (Equation 2.1) as a reference, it becomes obvious that companies have to have identical future expected financial benefits and identical cost of capital. The latter aspect typically means that companies have to exhibit the same systematic risk and equal capital structures. It is important to note that here it is abstracted from the possibility that companies have equal present values of future financial benefits but deviant future financial benefits and cost of capital. This abstraction is reasonable since, if such companies can be identified, there would be no need to apply CCV because the corporate values are already determined.

The immediate valuation model can also be put in formal terms. For reasons of better illustration it is not the general discounted cash flow model but rather the Gordon Growth model (Equation 2.2) that is used as a reference. Recalling that this model defines the corporate value as a function of the financial benefits, the cost of capital and the growth rate of financial benefits [$V_t = f\left(FB_\tau; c; g\right)$], we can derive the requirements necessary for the selection of comparable companies. The following simple equations describe the process of the immediate valuation approach of CCV:

$$V_{it} = V_{jt} \tag{2.7}$$

if $FB_{i\tau} = FB_{j\tau}$ and $c_i = c_j$ and $g_i = g_j$

with subscript i indicating the "target company"
subscript j indicating the "comparable company/companies"

Since the requirements to the comparability of companies are very high under this approach, it becomes obvious that its application typically fails because of the lack of equal companies in real settings. Thus, the aim of this section is not to provide a model for practical use but rather to highlight the importance of the principle of arbitrage. While this principle is strongly emphasized in the immediate valuation approach, it is not restricted to this model; rather it is the central element of every variant of CCV. However, the arbitrage principle no longer shows up in its pure form in the context of more complex CCV approaches; instead, some modifications are necessary to show how it dominates the models that rely on one or more reference variables.

---

39 Nevertheless, as could be learned from section 2.2, this parity only holds if investors' expectations are sufficiently uniform and valuation circumstances are identical. Thus, obviously, the arbitrage principle also serves as a demonstration of the requirements necessary to accurately perform CCV. For the further explanations in this section it should be assumed that these requirements are fulfilled.

2.3.2 Single-Factor Valuation Models

Mediate valuation models consider the fact that at real capital markets one can hardly find two or more companies that are perfect substitutes. Single-factor models use the relative difference in a typical financial key figure (e.g. earnings or book value) between the comparable companies and the target company to adjust the market prices of the comparable companies. In doing so, these models develop some kind of synthetic principle of arbitrage. More precisely, the prediction of these models is that, via the adjustment of the aggregated market prices of the peer group, a perfect substitute for the target company can be created. Since these substitutes should sell for the same price, the following equation must hold (see Peemöller et al., 2002: 197-198; Böcking and Nowak, 1999: 170; Ballwieser, 1991: 52; Nowak, 2000: 165; Wagner, 2005: 5-6):

$$V_{it} = V_{jt} \cdot \left[BR_{i\tau} \cdot \left(BR_{j\tau} \right)^{-1} \right] = \left[V_{jt} \cdot \left(BR_{j\tau} \right)^{-1} \right] \cdot BR_{i\tau} \tag{2.8}$$

The term $\left[V_{jt} \cdot \left(BR_{j\tau} \right)^{-1} \right]$ is often referred to as the "multiple", since it is the factor by which the basis of reference of the target company has to be multiplied to yield the value of the target company. It is important to note, that the multiple is calculated using data that are related to the comparable companies but not to the target company. Thus, the multiple can be seen as a market factor that is principally independent from the target company.

A major advantage of single-factor models compared to immediate valuation models is that the selection of comparable companies is less strict. Taking the Gordon growth model as the reference again and considering that only comparable companies are used for the determination of the multiple, the derivation of the multiple and thus the general requirements for the peer group selection can be stated as follows:

$$V_{it} \cdot \left(BR_{i\tau} \right)^{-1} = f\left(FB_{\tau} \cdot \left(BR_{\tau} \right)^{-1} ; c; g \right) \tag{2.9}$$

Obviously, the single-factor approach no longer requires the identity of all major value driving factors, but rather the identity of the cost of capital and the growth rate of financial benefits, as well as the identity of the ratio of financial benefits to the basis of reference. The latter relation is assumed to be equal over time, which dramatically simplifies the valuation process. Taking the PE ratio as an example, companies can be seen as comparable if they exhibit – along with equal cost of capital and growth rates – equal payout ratios. If it is further assumed that the "clean surplus relation" holds, the payout ratio can be completely eliminated as a selection criterion.[40] This is similarly true for other bases of reference. However, some of them have slightly higher, some lower requirements compared to the PE ratio.

The reduction of peer group selection requirements noticeably enlarges the number of possible comparable companies and therefore makes the single-factor models a valuation approach that is practically applicable. However, despite these obvious improvements compared to immediate valuation models, the remaining problems should not be treated lightly. While there are many companies with identical cost of capital, the sustainable growth rate – and therefore the future development of financial benefits – is not easily determinable in many cases. Thus, from an economic perspective, the identification of companies with identical growth rates is one of the big challenges for appraisers in single-factor CCV. For a more detailed examination of this matter, refer to sections 4.2.2.4 and 4.2.2.5, which emphasize the problems associated with the determination of the growth rate of financial benefits.

When composing the models it is important to define the numerator and the denominator consistently, i.e. if the variable in the numerator belongs to both capital providers (debt and equity providers), then the variable in the denominator has to belong to both capital providers, too; if the variable in the numerator only belongs to equity investors, the variable in the denominator should as well.[41] An example of this is the PE ratio where the price of a stock (the numerator) is a pure equity value and earnings (i.e. net income, the denominator) belong to equity investors as well – because claims of debt providers have already been satisfied when calculating net profit. Another example is the enterprise value/sales ratio (EV/sales ratio), where sales is an accounting figure that belongs to both debt and equity providers and the enterprise value is also a term for the market value of both equity and debt.

Typical bases of reference in single-factor models are the following accounting figures: net income (the respective model is the PE ratio), EBIT (earnings before interest and taxes; the respective model is the EV/EBIT ratio), EBITDA (earnings before interest, taxes, depreciation and amortisation; the respective model is the EV/EBITDA ratio), sales (the respective model is the EV/sales ratio), book value of equity (the respective model is the price-book ratio) and total assets (the respective model is the EV/total-assets ratio). Additionally, several kinds of cash flow can serve as reference variables.[42]

---

40 The clean surplus relation denotes the assumption that all changes in equity will be eventually included in capital contributions, dividends or earnings. See section 3.1.1 for more details about the similarity requirements of single-factor models and the clean surplus relation.

41 See Damodaran (2002: 457); Stowe et al. (2002: 225, 229-230); refer to section 3.2.1.2 for a more detailed presentation of differences between equity valuation and enterprise valuation.

42 The majority of all these models are described in-depth in Stowe et al. (2002: 179-237). In practical settings non-financial variables also sometimes serve as basis of reference, e.g. page views for valuing internet companies, or the number of customers

2.3.3 Multi-Factor Valuation Models

As in both of the models described above, the rationale behind the multi-factor CCV approach is again the principle of arbitrage. However, the derivation of an arbitrage-free basic equation is even more complex than in the single-factor approach. Multi-factor models are characterised by a number of bases of reference higher than one. As in the case of single-factor models these bases of reference are applied to adjust the value of the comparable companies so that a substitute for the target company's value is created synthetically. The multitude of factors, however, makes it difficult to modulate the corporate value of comparable companies simply by means of basic mathematical operations, as is the case in most single-factor models where the ratio of two variables is applied as adjustment mechanism. Therefore, more sophisticated and technically challenging approaches like, e.g., regression analysis or other combination approaches are usually appropriate.[43] The formal description of multi-factor models basically equals the general CCV equation presented in section 2.2.3.3 (Equation 2.4). For reasons of better illustration, it is rewritten here again, slightly modified to especially cover the multi-factor CCV model:

$$V_{it} = \sum_{n=2}^{x} a_{nt} \cdot BR_{jn\tau} \text{ resp. } V_{it} = a_{1t} + \sum_{n=2}^{x} a_{2nt} \cdot BR_{jn\tau} \quad (2.10)$$

It is important to note that the value of the comparable companies is included in the respective sensitivities a_{nt} or a_{2nt} (and in a_{1t}). The requirements for the comparable company selection cannot be determined across-the-board since they depend crucially on which of the three main value-driving factors $\left(FB_{\tau}; c; g\right)$ the model is able to eliminate. As in the case of single-factor valuation models, factors that are explicitly part of the valuation model drop out of the comparable company selection requirements. Consequently, the proceeding is straightforward: By putting more and more value explaining factors in the valuation model, the requirements become less and less strict. At the limit, if all value-explaining factors are part of the model, principally all companies can serve as comparable companies. For example, a multi-factor model explicitly using the financial benefits, the cost of capital and the growth rate as explaining factors would allow all properly priced companies to be part of the peer group. While theoretically applicable, such a model does not exist for one important reason: Appraisers are typically not able to accurately determine the long-term growth rate of financial benefits for all companies. If, however, they were able to determine the growth rate (and the other value driv-

---

for the valuation of telecommunication companies, see Coenenberg and Schultze (2002: 699); Schwetzler and Warfsmann (2005: 54-56). However, this thesis focuses on accounting reference variables.

43 An exception to this is the price-earnings-growth ratio (PEG) where the adjustment process proceeds by successively dividing the comparable companies' value by earnings and the growth rate; see Damodaran (2002: 487); Peemöller et al. (2002: 207); Schwetzler (2003: 81-82) and section 4.3.1.1.

ing factors as well) there would be no need for CCV because a direct valuation approach can be performed in a shorter time and – probably – with more accuracy.

Is there a way out of the growth rate determination problem? One possibility is to include several currently available factors – other than the long-term growth rate – in the model, in order to substitute the growth factor and the risk the company faces (see Damodaran, 2002: 463). This idea emerges especially in certain asset pricing models like the Fama-French multi-factor model (see Fama and French, 1992; Fama and French, 1993; Fama and French, 1996; Fama and French, 2000; Copeland et al., 2005: 873-875; for Germany see Ziegler et al., 2003) and certain variants of the arbitrage pricing theory.[44] These models are designed to be – and sometimes enjoy popularity as – a determination tool of the required return on investment and of the cost of equity capital in direct valuation approaches. However, there are also several models that aim to directly explain and determine market prices (see Ramakrishnan and Thomas, 1992; Barth et al., 1999; Herrmann, 2002: 118-120; for Germany see Möller and Schmidt, 1998: 477-504). From the perspective of CCV, there remains the risk that – due to the approach of empirically determining the factors and sensitivities with very big samples – such models lead to a prediction bias, since most likely not all companies in the sample are influenced by the same factors to the same degree. To put it differently, it cannot be ruled out that the predicted value might largely deviate from the real value of the target company for reasons of a cross-sectionally heterogeneous sample. However, accurate multi-factor CCV requires adequate accounting for the internal individuality of every single company and for the special external forces that influence the future development of this company.

To conclude: It remains to be shown whether there is a powerful model that includes all value-driving factors with no peer group selection necessary, or whether there is a more accurate model that applies some factors but leaves some other factors to the comparable company selection, or finally whether multi-factor models in general are not appropriate in CCV.

2.4 Special Tasks in Comparable Company Valuation

2.4.1 Quality of Accounting Variables

A crucial point in CCV is the quality of the accounting variables that serve as bases of reference. In this context, the term "quality" means to which degree the reported accounting figures are reasonable from an economic point of view. A major determinant of economic reasonability is the sustainability of accounting fig-

---

44 See Ross (1976); Burmeister et al. (1994); Copeland et al. (2005: 176-185). The major difference between the arbitrage pricing theory and other multi-factor models is that the arbitrage pricing theory is an equilibrium model, i.e. it is derived from an equilibrium theory.

ures (especially earnings).[45] When performing CCV, a high quality of accounting variables – meaning a high degree of similarity between reported and economicly reasonable figures – is preferable. This is the case since every theory-driven valuation approach requires that the major factors of the valuation process (here: the bases of reference) are not distorted but represent economic reality. Unfortunately, accounting standards usually allow the management of a company to select from a broad range of alternative acceptable accounting principles to describe certain economic circumstances. This might lead to a situation where the reported accounting figures differ (sometimes dramatically) from the economicly reasonable figures.[46]

A typical procedure to enhance the quality of accounting figures is to adjust them so that they equal or at least resemble the economicly reasonable figures (see Peemöller, 2003: 324; Palepu et al., 1996: chapter 3, 13; White et al., 2003: 2-3). However, two general problems are associated with that restatement process.

(1) The first problem is that there is no definite standard of when a variable is economicly reasonable. Different appraisers probably take different views on a certain issue even if they all look at the circumstances in a prudent way (see White et al., 2003: 4-5). However, there is certain evidence for when an accounting variable is not economicly reasonable. To put it more precisely, to overcome the problem appraisers typically rely on basic adjustment procedures. The following examples should offer some clarification.

To generate high quality earnings, it is often advised to ferret out and delete the non-recurring items and thus to enhance the sustainability of earnings. Such non-recurring items are, e.g., realized capital gains or losses, impairment or restructuring charges and charges due to catastrophes such as natural disasters or accidents (see White et al., 2003: 631-636; Cheridito and Hadewicz, 2001: 324-325). The book value of equity should be restated by increasing its closeness to the market value of equity. This could be done by using market values for all balance sheet accounts if market values are observable and available. It is, of course, virtually

---

45 See Friedlob and Schleifer (2003: 146); Penman (2001: 596). For financial statement analysis purposes the conservativeness of accounting figures is often seen as a determinant of accounting quality, too; see White et al. (2003: 637). However, this opinion does not apply for CCV since conservativeness often implies a downward bias of accounting numbers. Financial appraisers are rather interested in how well accounting describes the reality in an unbiased manner, see Palepu et al. (1996: chapter 3; 13); IDW (2002: 51-52).

46 For an in-depth discussion about accounting quality, see Penman (2001: 594-642); Palepu et al. (1996: chapter 3, 5-14); for a discussion about the problems of manipulating accounting figures in the context of valuation, see Gerke (2002:1); note: this section does not deal with the necessity for accounting choices from a pure accounting or auditing point of view. Neither does it deal with an assessment of different accounting principles. The main goal is to show which requirements the bases of reference have to fulfil so that CCV can be performed accurately. Additionally, most theoretical parts of this work are based on the assumption that accounting figures exhibit a very high quality in the sense of CCV.

impossible to revalue all items. However, some progress can be made in this direction.

Contrary to a widespread belief, accounting differences also have an impact on cash flows. There is a direct influence because the amount of taxes paid – which is an outflow of cash – crucially depends on the accounting choices of a company.[47] Moreover, an indirect influence also exists because accounting choices affect the classification of different cash flows. For example, if leases are classified as operating leases under US-GAAP the whole lease payments are regarded as operating cash flow. If they are classified as capital leases then the repayment of the debt component of the lease payments is regarded as financing cash flow. Differences in cash flow classifications also arise depending on whether items are capitalised or expensed, e.g. if expensing is chosen for an item then operating cash flow is more negatively affected. From the point of view of CCV, no adjustments are necessary for the taxes paid because they are real cash outflows. As long as free cash flows serve as a basis of reference, reclassifications are not necessary either (see Penman, 2001: 606. As regards the problems of using free cash flows as a basis of reference, see section 4.2.1). This is the case because the calculation scheme for free cash flows does not follow the accounting scheme. Contrary to this, if operating cash flows serve as a basis of reference, then some adjustments might be advisable.

It is important to note that all adjustments have to be done consistently for the target company as well as for the comparable companies. In this context, it becomes obvious that the use of different accounting principles (IAS/IFRS, US-GAAP, German GAAP, etc.) for different comparable companies or the target company clearly hampers the restatement of accounting figures. Since German GAAP is a rather creditor-oriented system and therefore many adjustments are necessary to create (equity-) investor-oriented figures, the German Society of Investment Appraisers and Asset Managers (DVFA) established a system on how to adjust earnings (see Busse von Colbe, 2000). This scheme is often used in CCVs (see Löhnert and Böckmann, 2005: 411).

(2) The second problem is that a lack of data complicates the adjustment process. This is a general problem in financial statement analysis. In most cases it concerns off-balance sheet adjustments. For example, the reclassification of operating leases to capital leases typically requires the appraiser to estimate certain value determining factors that are not disclosed. Additionally, as mentioned above, it concerns some already recorded items, too: The revaluation of single assets is only possible as long as market values are observable and available (see White et al., 2003: 246-248, 621; Bernstein and Wild, 1998: 175-178, 184, 189-

47 Certainly, taxes payable are determined based on tax accounting rules. However, in Germany the tax and financial accounts are interdependent due to the so-called "authoritative principle" (steuerliches Maßgeblichkeitsprinzip). This means that financial statement accounting rules (at least those applied on an unconsolidated basis, i.e. German GAAP) largely determine the tax accounting. As a consequence, financial accounting choices directly affect the amount of taxes payable.

190, 192-193). The availability of data may differ for different accounting standards.

After the basic adjustment operations are conducted, accounting variables can principally serve as an economicly reasonable basis of reference. However, as outlined above, there is still room for accounting discretion and consequently for different interpretations of economic reality. A phenomenon that is often experienced, even after the basic adjustments, is that companies with similar operations have very different reported figures due to differences still remaining in accounting choices. Typically, problems arise when dealing with last-in-first-out (LIFO) versus first-in-first-out (FIFO) inventory valuation methods, the depreciation methodology as well as the determination of the useful life, the methods used to account for intercorporate investments (cost, equity or consolidation method), the employee stock option accounting, the retirement benefit accounting and other management accounting choices such as the timely recognition of sales and cost or the use of provisions (see White et al., 2003: 621-628, 631-636). To better handle that problem in CCV, appraisers have to adhere to one important rule: The accounting variables that serve as bases of reference not only have to be adjusted so that they represent economic reality, they also have to be adjusted so that they represent economic reality in an identical manner for all firms. This means that adjustments have to be done consistently for all companies that are part of the valuation process (see Löhnert and Böckmann, 2005: 416; ASA, 2002: 7, 32).

The outlines of accounting quality, thus far, can only offer a rough insight into the accounting facets of CCV. However, the focus of this work is clearly on the economic aspects of CCV. Therefore, no in-depth discussion about accounting aspects is conducted. Moreover, the assumption of perfect quality of accounting variables underlies the theoretical part throughout (chapter 2, 3 and 4).[48]

2.4.2 Aggregating the Peer Group Results

Another important issue in CCV is how to aggregate the peer group results if there is more than one comparable company. There are typically four different methods of putting together the data of comparable companies in order to create the substitute for the target company: (1) determining the arithmetic mean of the multiples, (2) the median of the multiples, (3) the harmonic mean of the multiples or – last, but not least – (4) applying a regression approach. The arithmetic mean of the multiples is simply the average of the computed multiples. It is calculated as the sum of all the observed outcomes divided by the total number of observations (see Berck and Sydsæter, 2000: 27). In the context of single-factor comparable company valuation, this leads to the following aggregation equation:

[48] For a more detailed discussion of the accounting challenges in the context of financial statement analysis and CCV, see White et al. (2003); Bernstein and Wild (1998); Peemöller (2003); Krolle et al. (2005).

$$\left[V \cdot (BR)^{-1}\right]_{aggregated} = \sum_{x=1}^{m}\left[V_x \cdot (BR_x)^{-1}\right] \cdot m^{-1} \tag{2.11}$$

with m being the number of comparable companies

A problem using the arithmetic mean is that if single data points are very far away from the rest of the multiples (outliers), the mean will then be strongly influenced by these outliers. In such cases, the median is a better measure because it is quite resistant to such outliers. In the case of CCV, the median is either the middle score of the multiples ranked in ascending order (for an uneven number of multiples) or the average of two middles (for an even number of multiples):

$$\left[V \cdot (BR)^{-1}\right]_{aggregated} = \begin{cases} \left(V \cdot (BR)^{-1}\right)_{(m+1)/2} & \text{for uneven } m \\ \left[\left(V \cdot (BR)^{-1}\right)_{m/2} + \left(V \cdot (BR)^{-1}\right)_{m/2+1}\right] / 2 & \text{for even } m \end{cases} \tag{2.12}$$

for multiples $\left[V \cdot (BR)^{-1}\right]$ that are ranked in order of value.

The third aggregation method is the harmonic mean. It is computed as the reciprocal of the average of the reciprocals of the multiples (see Berck and Sydsæter, 2000: 27):

$$\left[V \cdot (BR)^{-1}\right]_{aggregated} = m \cdot \left(\sum_{x=1}^{m}\left[V_x \cdot (BR_x)^{-1}\right]^{-1}\right)^{-1} \tag{2.13}$$

Effectively, it is the multiple of a portfolio, which consists of the same amount of funds invested in terms of the basis of reference in every comparable company (see Pratt, 2001: 133).

Finally, the regression approach aggregates the multiples by determining a line of best fit. This line is such that it minimises the squares of the residuals of all observations. An example for this approach can be found in section 4.3.2. An advantage of the regression approach is that it allows a reasonable calculation of corporate values if the basis of reference is zero or even negative (if the regression model does not restrict the intercept to be zero).[49]

The arithmetic mean is the most popular of all the aggregation methods outlined above. This is largely due to its simplicity. It is a reasonable approach especially if the computed multiples of the individual comparable companies are not very dispersed or if this dispersion follows a normal distribution. While it can be verified easily without any in-depth analysis as to whether the first condition

49 Another aggregation method is the value weighted mean, see Herrmann (2002: 105). However, because of its low practical relevance it is not discussed in details here.

holds, the existence of normal distribution is rather difficult to verify because of the typically small number of comparable companies.[50] Additionally, in small samples single outliers can dramatically change the degree of skewness of the distribution. Therefore, it is more appropriate and economicly sound to use the median in any case, if the multiples of the comparable companies are dispersed (see Damodaran, 2002: 459-460; Schwetzler, 2003: 88). Furthermore, rule of thumb advises appraisers to use the median rather than the arithmetic mean if the sample of the comparable companies is very small (equal to or less than 4 companies as a rule).

Often, and for a number of reasons, the harmonic mean is denoted a very attractive alternative in peer group aggregations from a theoretical point of view (see Schwetzler, 2003: 88-89; Pratt, 2001: 133). First, the harmonic mean is in accord with the fact that the cost of equity (which are negatively correlated to the corporate values and the respective multiples and therefore their behaviour is basically similar to the reciprocals of the multiples) are additive in the CAPM world (see Schwetzler, 2003: 89). Second, it is a more conservative measure of central tendency than the arithmetic mean and does not suffer from the upward bias that is associated with other aggregation methods (see Baker and Ruback, 1999: 16; Beatty et al., 1999: 182). Finally, the use of the harmonic mean is also supported by the results of some empirical studies, which compare the pricing accuracy of multiples using different aggregation variants (see Baker and Ruback, 1999: 16, 20; Beatty et al., 1999: 182; Liu et al., 2002: 22). A problem with the harmonic mean is, however, that it consequently overweights the low priced comparable companies, which might not be in accordance with economic reality. For all three aggregation methods discussed thus far it is important to consider that, if there are negative-earnings companies as part of the sample, the aggregation has to be done at the accounting variable level and not at the multiples level. Failing to do this might lead to the computation of biased aggregated multiples (see Benninga and Sarig, 1997: 314-317).

The regression approach is a reasonable approach only if the sample of comparable companies is large enough to allow accurate results. Additionally, because of its relative complexity compared to the other methods, it is typically only used if the intercept is intentionally not restricted to zero or if the valuation model comprises more than one factor (see Meitner, 2003; Meitner, 2004).

Figure 2 visualises the impact of different aggregation methods on the determination of the PE ratio. The PE ratio is the slope of the respective aggregation line. However, the results presented here for a fictitious set of companies *A* to *K* are not restricted to the PE ratio. They are principally the same for all single-factor models and also some of the multi-factor models. In the figure, the PE ratios are indicated in brackets for every single company. Company B, for example, has a PE ratio of 27.8 which would also be the tangent of the grey coloured angle α (=tan α) in the origin, assuming that the two axes are equally scaled. The general principle

[50] This is a problem because the "central limit theorem", which postulates that the probability of data being normally distributed increases with the number of observations, cannot be applied here; see Hogg and Craig (1995: 246-253).

is: The bigger the angle in the origin for a company, the higher the PE ratio. Obviously, the sample suffers from two outliers (company *A* and *B*) which both have a very steep slope seen from the origin. As a result, the sample median noticeably deviates from the arithmetic mean. The harmonic mean lies between the arithmetic mean and the median. Unlike the median it considers the existence of the outliers, but unlike the arithmetic mean it does not weight them that strongly. The regression approach delivers totally different valuation results, especially in – but not limited to – the low earnings brackets. Its PE-line does not include the origin and the slope is much flatter.

Fig. 2. PE ratio for different aggregation methods

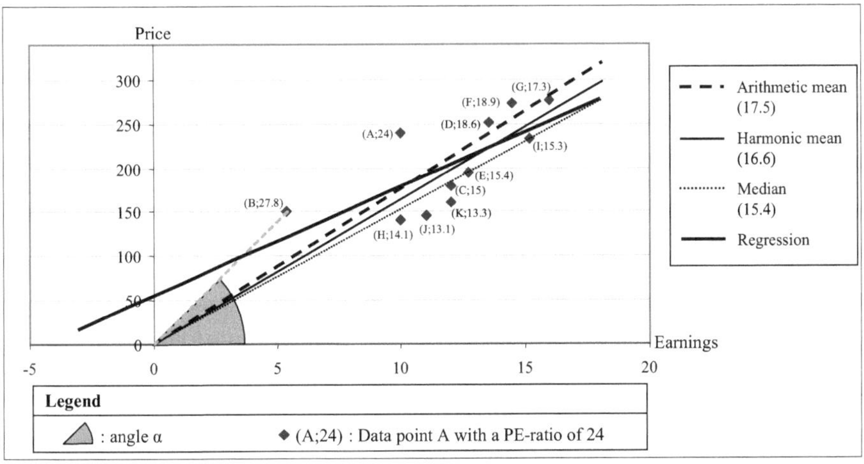

Source: ZEW

Regardless of the technical aspects of putting together the comparable companies' data, it is important to note that the aggregation problem only exists if the comparable companies are not perfectly similar as required by the valuation model and/or the factual degree of capital market efficiency allows for mispricing of single companies. Actually, both situations more or less mirror the circumstances in real valuation settings. Additionally, certain valuation models require the use of a special aggregation method. Thus, from an economic point of view, it is necessary to have a closer look at the circumstances surrounding the valuation process to identify the benefits and applicability of each particular aggregation approach in the context of CCV. This will be done in the remainder of this section.

Provision for the degree of similarity of comparable companies

In some cases appraisers have to rely on a more or less heterogeneous group of comparable companies. Usually, CCV requires them to include only companies that meet the high requirements of similarity but in some valuation settings such companies are simply not available. Companies can only be put into the peer group if they do not deviate too much from the target company. Additionally, it is appreciable that these companies *on average* equal the target company. That means that none of the value-driving variables is different in one specific direction between both groups of companies.[51] Nevertheless, the more heterogeneous the sample is, the more accurate it is to use the median instead of the arithmetic mean because the risk of the existence of outliers grows with rising heterogeneity. Applying the harmonic mean instead of the arithmetic mean is also more reasonable if companies tend to be heterogeneous. Furthermore, the regression approach is no longer appropriate if the sample is not homogeneous. Finally, the choice between harmonic mean and median is not affected by the degree of similarity of the peer group companies.

Provision for factual market efficiency

The degree of market efficiency also determines the aggregation method for the peer group multiples. Market efficiency is a measure of the extent to which information is reflected in prices. Even in the strongest form of market efficiency, not every company is necessarily priced correctly but all companies are on average priced correctly, which means that companies' multiples are either tightly clustered and/or any mispricing is normally distributed around the intrinsic value (see Fama, 1976: 144; Damodaran, 2002: 113; Eidel, 2000: 13). With decreasing degree of market efficiency, however, the risk of systematic mispricing increases. This, in turn, increases the probability of the peer group sample being affected by outliers.[52] Thus, the more the relevant market is inefficient the more are appraisers advised to apply the harmonic mean or the median instead of the arithmetic mean. It is also reasonable to set the application of the regressional approach aside if there is an observed skewness in the sample. Again, market efficiency does not influence the choice between harmonic mean and median.

Provision for the applied valuation model

While single-factor models typically allow for all aggregation methods outlined above, some multi-factor models do not. Certainly, multi-factor models like the PEG ratio or combinations of single-factor models are principally open to any aggregation mechanism. However, multi-factor models that rely on regression based factor sensitivities are restricted to use the regression approach in peer group aggregation, too. The derivation and discussion of such a model is presented in section 4.3.2.

2.4.3 Premiums and Discounts

In some cases the choice of the appropriate aggregation method cannot totally remedy the differences between the peer group and the target company. Additionally, sometimes the purpose of valuation is such that market prices are not an adequate basis for determining the corporate value. In such situations, the use of pre-

---

51 See section 3.1.1 for an in-depth analysis of the similarity of companies.

52 See section 3.1.2 for an in-depth analysis of the influence of market efficiency on the CCV process.

miums and discounts is a way of adjusting the computed value of the target company to yield better valuation results.

Analysts and investors regard the determination of appropriate premiums and discounts as a very difficult task in CCV. A survey among 203 financial analysts and institutional investors, conducted as part of the monthly financial market survey of the Centre for European Economic Research (ZEW) (see Beckmann et al., 2003), reveals that almost 60 % of all respondents find it problematic to ex post adjust the result of CCV (answer categories 4 and 5). In contrast, only 13 % do not expect any major problems arising from this procedure (answer categories 1 and 2) (see Beckmann et al., 2003: 104-105). This is highlighted in Figure 3.

Fig. 3. Premiums and discounts in the eyes of analysts and investors

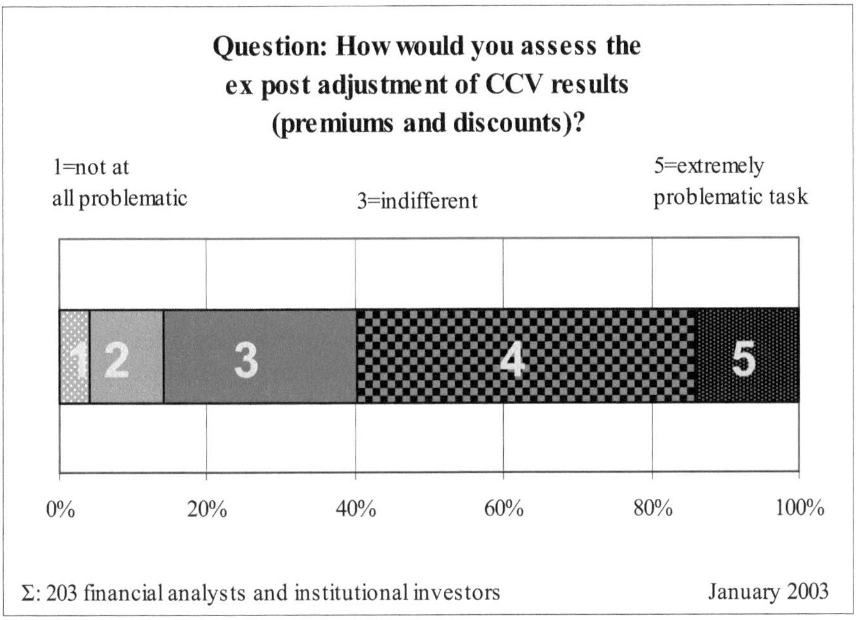

Source: ZEW

Below, some circumstances are discussed where the use of premiums and discounts is appropriate. However, it is important to note that such adjustments are a very delicate issue. They require a great deal of subjectivity and therefore provide room for misvaluations and even manipulations. CCV should rather be declared failed if there is a lack of comparable companies, than be performed under the use of overly discretionary premiums or discounts.

Adjustments due to existing differences between peer group and target

Sometimes there are systematic minor differences in the nature of the peer group and the target company, resulting in a higher value of the target than of the average peer group company. In such situations it might be appropriate to adjust the appraisal value (that is determined based on CCV) of the target upward by adding a premium. Such differences exist, e.g., if a company is more profitable than the comparable companies or if it has certain competitive advantages that other peer group companies do not have. Contrary to this, if a company performs principally worse than the rest of the peer group or has a weaker strategic and competitive position, then the application of a discount would probably be appropriate (see Peemöller et al., 2002: 206).

However, because of the high level of subjectivity that is associated with such adjustments, appraisers should be very careful in applying premiums and discounts due to differences between the peer group and the target (see Hillebrandt, 2001: 619). Moreover, in such a situation it is very important to assess whether it is still justified to call the peer group "comparable companies". In fact, in some cases the differences between target and peer group are too big and, thus, a reasonable CCV is no longer possible.[53]

Adjustments due to existing differences between price and value

Even if there are no systematic differences between the group of comparable companies and the target company, there might be demand for value adjustments. A typical application of premiums and discounts is the situation in which the corporate value deviates from the price of the company. Bearing in mind that prices determined at stock markets typically reflect the value of minority shares, there might be a need for upward adjustments if the appraiser is interested in the value of a controlling interest – this adjustment is called the control premium (see Pratt, 2001: 136-144). To determine the control premium appraisers often have to rely on historical data. Based on such data, premiums are determined by comparing the price paid in an acquisition of a controlling interest compared with the public market trading price just prior to the announcement of that acquisition. Studies for the US stock market show a yearly premiums' median of about 30-35 % and an unweighted arithmetic mean of about 40 % (see Pratt, 2001: 139; Hanouna et al., 2001; Gaughan, 1999: 520). In Europe, premiums are slightly lower. They average at about 30 %.[54] It is, however, important to note that control premiums are not necessarily stable over time but are rather positively related to the general level of stock prices (see Schwenker, 2001: 9). For example, at the top of the stock market boom average acquisition premiums rose up to 50 % (September 2000) (see Heller, 2001. Meanwhile, premiums of most transactions have returned to moderate levels, see Ecker and Ehren, 2004: 89).

Consequently, the empirical evidence can be only a rough indicator for the height of the premium that should currently be included in CCVs. Appraisers need to have a good sense of how capital markets work, in order to accurately adjust market prices so that they can determine the value of a controlling interest.

---

[53] See section 3.1.1 for a discussion about the similarity requirements of comparable companies depending on the applied valuation model.

[54] See Schwenker (2001: 9). Franks et al. (1988) reported an average premium of about 30 % for cash-financed deals and of about 15 % for stock-financed deals in the UK-market.

Literature sometimes names the opposite of a control premium – a minority discount – as another adjustment mechanism, which might be appropriate in CCV (see Pratt, 2001: 136-137). However, as far as the peer group consists of publicly listed companies this variant is of no paramount importance in real valuation settings. Consequently, no stress is put on that issue here.

Adjustments depending on the market structure

Another case of common usage of premiums and discounts is that of differences in the stock market liquidity of the target company and the comparable company. Such a difference typically arises when valuing private companies via the market approach (i.e. the shares of the target cannot be freely traded in a timely manner). However, it can also arise when the market allows the shares of the target company to only be thinly traded. This lack of liquidity and/or marketability of the target company typically results in a corporate value that is lower than it would be if shares are publicly traded especially because of higher transaction and opportunity costs. Consequently, appraisers have to downward adjust the value computed by CCV in order to yield a more realistic corporate value (discount for lack of marketability or discount for lack of liquidity) (see Pratt, 2001: 145-155). Empirical studies found that these discounts are substantial in size. Examinations that compare restricted stocks with publicly traded stocks of the same company revealed an average discount of about 30-35 % for the US stock market (see Pratt et al., 2000: 404; Nowak, 2000: 168).

It is important to note that there is no formal link between control premiums and marketability discounts. Both phenomena can arise independently from each other and therefore have to be analysed independently from each other (see Booth, 2001: 5-6).

3 Interrelation of Comparable Company Selection and Valuation Model Choice

The two core-tasks of CCV are the selection of comparable companies and the choice of an appropriate valuation model. While the first one deals with composing the sample of companies, the second involves choosing between single-factor and multi-factor models as well as selecting the appropriate reference variables. A survey among 203 financial analysts and institutional investors, conducted as part of the ZEW-financial market survey in 2003 (see Beckmann et al., 2003), revealed that both procedures are challenging tasks. More than 40 % of the interviewees regard the peer group selection as problematic (answer categories 4 and 5), while even 50 % admitted that the valuation model choice is associated with problems (answer categories 4 and 5). The percentage of those who see both tasks as rather uncomplicated assignments (answer categories 1 and 2) is only 21 with regard to the comparable company selection, resp. 20 for the valuation model choice (see Beckmann et al., 2003: 104). A graphical depiction of these results is provided in Figures 4 and 5.

Some authors describe comparable company selection and valuation model choice as separate tasks. In most cases they recommend starting with the company selection, then switching over to the valuation model choice.[55] However, comparable company selection and valuation model choice are in no way two independent steps of the CCV approach, but rather closely connected tasks (see Herrmann, 2002: 99). It is not possible to appropriately carry out one of these tasks without considering the other task. More precisely, there is a strong interrelation between the two main parts of the CCV process, which has to be considered by appraisers. This interrelation, in turn, makes CCV an ambitious valuation approach with high economic requirements to be met.

The remainder of this section involves a discussion of both comparable company selection and valuation model choice, with the focus on a detailed description of the determinants that influence the handling of each of these two tasks. Based on that, the section provides insights into the interaction between peer

---

[55] See e.g. Barthel (1996: 156); Peemöller et al. (2002: 203-205); Schmidbauer (2004: 150); Küting and Eidel (1999: 229); Böcking and Nowak (1999: 171). Löhnert and Böckmann (2005: 411) recommend starting with the model selection and composing the peer group afterwards.

Fig. 4. Comparable company selection (in the eyes of analysts and investors)

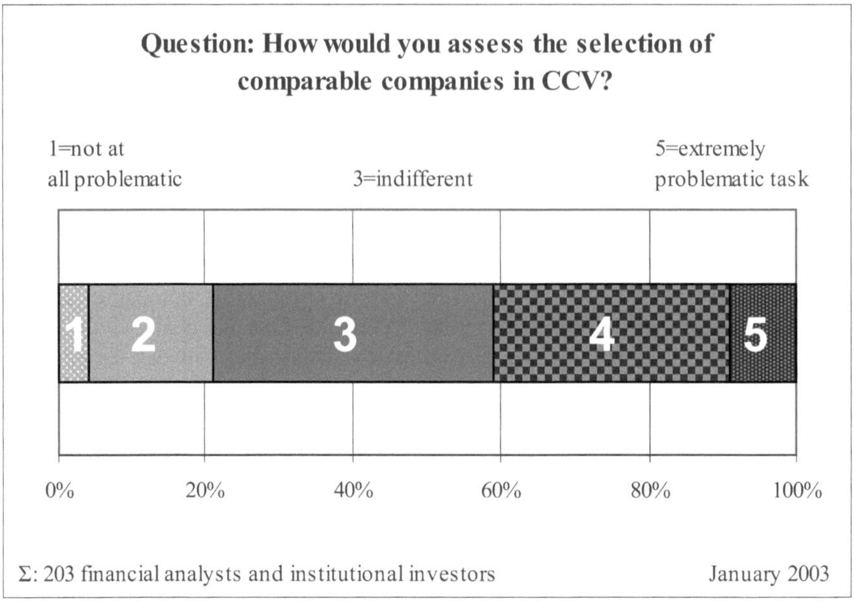

Source: ZEW

Fig. 5. Valuation model choice (in the eyes of analysts and investors)

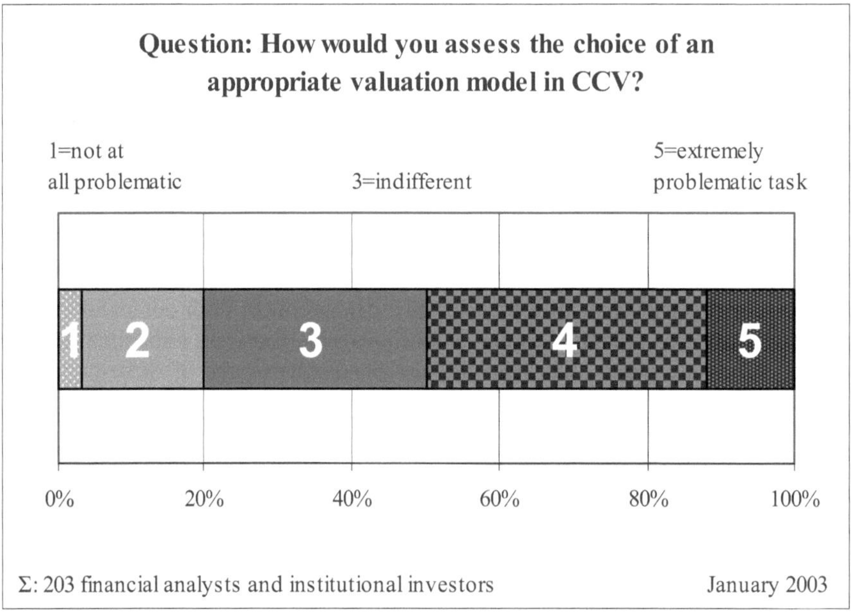

Source: ZEW

group selection and valuation model choice. Special emphasis is put on how decisions in one task affect the decisions in the other task. The section closes with guidance on how the two core-parts of CCV should be performed given various different external circumstances.

3.1 Determinants of Comparable Company Selection

The selection of comparable companies is mainly driven by two external determinants. The first one is the degree of similarity between the peer group companies and the target company. This factor is a function of the availability of comparable companies in real valuation settings. The second determinant is the degree of efficiency and pricing quality in the capital market relevant for the valuation of the target company.

3.1.1 Degree of Similarity of the Peer Group Companies

All articles, books and academic papers about CCV agree that comparable companies must have similar characteristics to the target company. However, results are mixed regarding which specific characteristics should be concordant. In particular, two different general approaches exist to assess this similarity. The first one is a theory-driven approach based on valuation model implications, while the second one is a rather practice-oriented non-technical approach.

Valuation theory driven approach to assessing the similarity of companies
In the theory-driven approach the selection of peer group companies is attributed to the basic idea of the Gordon growth model (Equation 2.2). By comparing CCV models with this direct valuation model, it becomes clear which characteristics companies should exhibit to serve as comparable companies in CCV. For single-factor models this is indicated by the general Equation 2.9. However, dependent on the specific CCV model, the concrete determinants of similarity might vary due to technical rearrangement of the respective valuation equation and changing economic consequences. A more detailed description of the link between the Gordon growth model and selected single-factor models should clarify the similarity requirements prescribed by that approach.[56]

---

[56] See Damodaran (2001a: 251-352); Damodaran (2002: 460-462, 470-472, 514-515 and 545-547); Richter (2002: 53). Similar analysis is provided by Herrmann (2002: 98-100, 125-184); Richter and Timmreck (2000: 279) and Moser and Auge-Dickhut (2003: 215-218). The latter three sources provide a detailed analysis of the fundamental determinants of certain single-factor models. Their approaches especially highlight the importance of multiple different growth stages and are therefore far more complex in nature than the approach provided here. However, for the determination of the similarity criteria it is sufficient – and even more target-oriented – to rely on the basic

The right hand side of Equation 3.1 highlights the determinants for the required similarity of comparable companies when using the trailing PE ratio. This equation is the result of a simple rearrangement of Equation 2.2.[57]

$$PRICE_t \cdot \left(NI_t\right)^{-1} = \left(DIV_t \cdot \left(NI_t\right)^{-1}\right) \cdot \left(1+g\right) \cdot \left(c_e - g\right)^{-1} \qquad (3.1)$$

Thus far, the similarity requirements are: identical payout ratio, identical growth rate and identical cost of equity. Additionally assuming that the clean surplus relation holds, however, eliminates the influence the payout ratio has[58] and leads to the final similarity requirements:

$$PRICE_t \cdot \left(NI_t\right)^{-1} = f\left(c_e; g\right)$$

with $\Delta PRICE / \Delta c_e < 0$ and $\Delta PRICE / \Delta g > 0$

The elasticities show how minor differences in one similarity criterion of the target company compared to the comparable companies affect the price of the target company.

Equation 3.2, which is also called the Gordon/Shapiro equation (see Beaver and Morse, 1978: 65; Damodaran, 2002: 471), see also Equation 2.5, shows the functional relationship for the (one year ahead) leading PE ratio:

$$PRICE_t \cdot \left(NI_{t+1}\right)^{-1} = \left(DIV_{t+1} \cdot \left(NI_{t+1}\right)^{-1}\right) \cdot \left(c_e - g\right)^{-1} \qquad (3.2)$$

Again, the impact of the payout ratio can be eliminated as a comparability criterion by assuming a clean surplus relation. It becomes obvious that, except from a different intensity of the influence of the growth rate, the requirements are identical to that of the trailing PE ratio:

Gordon growth model. This will become clearer when the economic determinants of corporate growth are analysed more in- depth in section 4.2.2.4 and 4.2.2.5.

57 Note that the corporate value is now substituted by the stock price, since no differences are assumed between equity value and market price for the formal analyses.

58 See Lundholm (1995: 752). The clean surplus relation means that only capital contributions, dividends and the profit or loss reported in the income statement can change the amount of owners' equity. To put it differently, all revenues and expenses have to be recorded on the income statement. For IAS/IFRS, US-GAAP and German GAAP the clean surplus relation only holds approximately. That in turn means that all three accounting regimes slightly tolerate a "dirty" surplus accounting. In this context, it has been shown that the accounting does not necessarily have to be clean surplus in order for the payout ratio being irrelevant, see Yee (2005). To conclude, the payout ratio is possibly not totally meaningless for the PE ratio. However, it can be assumed that its importance is limited.

$$PRICE_t \cdot (NI_t)^{-1} = f(c_e; g)$$

with $\Delta PRICE / \Delta c_e < 0$ and $\Delta PRICE / \Delta g > 0$

The differences in the impact of the growth rate are due to the lower price elasticity of the growth rate in the case of the leading PE ratio compared to the trailing PE ratio. This makes the leading PE ratio less sensitive to differences in the growth rate between the target company and the comparable companies than the trailing PE ratio.

Proceeding in the same way for the price book (PB) ratio reveals the following relationship (see Damodaran, 2002: 514-515; Kuhlmann, 2005: 97-98):

$$PRICE_t \cdot (EQUT_t)^{-1} = \left(DIV_{t+1} \cdot (NI_{t+1})^{-1} \right) \cdot ROE_{t+1} \cdot (c_e - g)^{-1} \tag{3.3}$$

with $ROE_{t+1} = NI_{t+1} \cdot (EQUT_t)^{-1}$

Under the assumption of the existence of the clean surplus relation – which again leads to the elimination of the payout ratio – the similarity requirements are as follows:

$$PRICE_t \cdot (EQUT_t)^{-1} = f(ROE; c_e; g)$$

with $\Delta PRICE / \Delta ROE > 0$ and $\Delta PRICE / \Delta c_e < 0$ and $\Delta PRICE / \Delta g > 0$

Table 1 exhibits the functional relationship, the similarity requirements and the elasticities for several other commonly used single-factor models. As can be seen, the higher the position of the reference variable in the income statement, the higher the requirements for the comparable company selection. While the PE ratio only requires the growth rate and cost of equity to be similar, the EV/EBIT ratio additionally requires the EBIT-earnings-margin – and therefore primarily the financing activities – to show a resemblance. The EV/SALES ratio even requires the profit margin to be alike (see also Moser and Auge-Dickhut, 2003: 20; Damodaran, 2002: 461; Herrmann, 2002: 52-53, 99).

Table 1. Similarity requirements for selected single-factor models

Reference variable	Functional relationship	Elasticities
Cash flow	$PRICE_t \cdot (CFO_t)^{-1} = (1+g) \cdot (c_e - g)^{-1}$	$\Delta PRICE / \Delta c_e < 0$ $\Delta PRICE / \Delta g > 0$
	similarity requirements:	c_e, g
EBIT[a]	$EV_t \cdot (EBIT_t)^{-1} =$ $= NI_t \cdot (EBIT_t)^{-1} \cdot \left[(1+g) \cdot (wacc - g)^{-1}\right]$	$\Delta EV / \Delta wacc < 0$ $\Delta EV / \Delta g > 0$ $\Delta EV / \Delta(NI_t / (EBIT_t)) > 0$
	similarity requirements:	$wacc$, g, $NI_t / (EBIT_t)$
EBITDA[a]	$EV_t \cdot (EBITDA_t)^{-1} =$ $= NI_t \cdot (EBITDA_t)^{-1} \cdot \left[(1+g) \cdot (wacc - g)^{-1}\right]$	$\Delta EV / \Delta wacc < 0$ $\Delta EV / \Delta g > 0$ $\Delta EV / \Delta(NI_t / (EBITDA_t)) > 0$
	similarity requirements:	$wacc$, g, $NI_t / (EBITDA_t)$
SALES[a,b]	$EV_t \cdot (SALES_t)^{-1} =$ $= X_t \cdot (SALES_t)^{-1} \cdot \left[(1+g) \cdot (wacc - g)^{-1}\right]$	$\Delta EV / \Delta wacc < 0$ $\Delta EV / \Delta g > 0$ $\Delta EV / \Delta(NI_t / (SALES_t)) > 0$
	similarity requirements:	$wacc$, g, $NI_t / (SALES_t)$

EV= Enterprise Value, CFO = Operating Cash Flow, EBIT = Earnings before Interest and Taxes, EBITDA = Earnings before Interest, Taxes, Depreciation and Amortization, wacc = weighted average cost of capital.

[a] These reference variables are determined before considering the financing of the company, i.e. before interests. Consequently they belong to all capital providers, debt and equity holders, and therefore they have to be set in relation to the value of the company as a whole: the enterprise value *EV*. Note: In all these cases it is assumed that the clean surplus relation holds. [b] Sometimes appraisers apply a price/sales ratio in CCV. This, however, is inconsistent with valuation theory since sales are determined before financing and therefore have to be set in relation to the value of all capital providers, which is the enterprise value *EV*. A thorough discussion about this issue can be found in section 3.2.1.2.

Sources: Damodaran, 2002: 460-462, 470-472, 514-515 and 545-547, ZEW.

Practice-oriented approach to assessing the similarity of companies

Contrary to the first approach, this approach aims to determine the group of comparable companies by simply looking for similarities in the corporate structure, strategic position or the business model of the company. These similarities correspond (but are not limited) to the following criteria (see Barthel, 1996: 150; Bausch, 2000: 455; Cheridito and Hadewicz, 2001: 322; Löhnert and Böckmann, 2005: 414; Peemöller et al., 2002: 204-205; Schmidbauer, 2004: 150; Damodaran, 2001a: 266; Böcking and Nowak, 1999: 171):

> Products or services offered, depth of the value added process, industry and type of business, share in key markets, technological level of products (low end, mid range, high end technology), technological level of infrastructure, distribution channels and type of customers, position in and length of product life cycle, degree of competition in the respective industry, influence of competitors, degree of product diversification, legal environment, tax situation of the company and its shareholders, regional revenue distribution, quality and experience of management, size of the company, legal form of the company, capital market orientation (public or private company), capital structure, operational risk (business risk).

Considering the limited number of public companies in reality, it becomes obvious that it is difficult to find firms that perfectly match in all of these criteria. However, some possibilities exist to deal with that problem in practice.

(1) Many of the criteria cited above can be substituted by simply considering the "industry" criterion (see Löhnert and Böckmann, 2005: 414-415; Wagner, 2005: 14). Usually, all companies in one industry face the same depth of the value added process, the same position in and length of the product life cycle, identical degree of competition, identical competitors etc. Of course, there are also certain criteria that cannot be summarised by the industry criterion. These are, e.g., the market share, the size of the company and the capital structure. Nevertheless, to include the industry as a selection criterion dramatically simplifies the composition of the peer group.

(2) Not all criteria are equally weighted. Some are more important than others and therefore the similarity in these criteria is more highly appreciated (see Cheridito and Hadewicz, 2001: 322). Except for the accepted dominating role of "industry classification", however, guidelines offer little advice as to which of the characteristics listed above is of higher importance and which is of lower importance in the process of selecting comparable companies.

(3) Similarity does not necessarily mean that the companies must be identical but rather that relevant criteria should be "alike".[59] Consequently, appraisers always have some sort of discretion when composing the peer group. However, ap-

[59] In this context Hickmann and Petry stated: "While the comparable firms are not identical, they are similar and should therefore have similar prices", Hickmann and Petry (1990: 77); see also Nowak (2000: 16); ASA (2002: 12).

praisers are advised to be very careful with that discretion. In case of doubt whether a company deviates too much from the target company in one criterion, that company should be excluded from the peer group rather than included.

Comparison of both approaches

At first glance these two approaches seem to be highly different. However, a closer look on the fundamental idea of each approach reveals the connection between them. Obviously, the theory-driven approach is economicly reasonable. By using accredited valuation models as a reference – and therefore linking CCV to direct valuation methods – this approach provides a closed valuation framework. Assuming that the Gordon growth model is a theoretically sound valuation model inevitably leads to the conclusion that the valuation theory-driven approach is theoretically sound, too. The limit of the practical applicability of this approach, however, lies in the imperfections of real valuation settings. Taking into account that individuals typically have a limited forecasting ability, it becomes obvious that the determination of some of the similarity characteristics is a difficult and sometimes even an unrealisable task. To call a spade a spade: The main problem is the accurate specification of the long-term growth rate since it requires appraisers to forecast the future cash flows of each company for a longer time horizon.[60]

That is where the practice-oriented approach comes into play. By relying on directly observable, readily available similarity characteristics, the forecasting problem is totally eliminated. Most of these observable characteristics should serve as a substitute for the long-term growth rate (see Peemöller et al., 2002: 204). The main idea behind that approach is that companies with certain identical observable characteristics should have similar long-term growth rates.

However, the use of the practice-oriented approach is a double-edged sword. On the one hand, it is important to select only those characteristics that reasonably contribute to the substitution of the growth rate because otherwise the number of comparable companies would be unnecessarily reduced. On the other hand, the appraiser should ensure that they include all characteristics that serve as a substitute for the long-term growth rate since otherwise the sample may possibly include companies that are dissimilar to the target company. As a consequence, the practice-oriented approach is possibly but not necessarily a reasonable comparable company selection mechanism. Since there is neither empirical evidence that could serve as a benchmark nor any publicly accepted, self-contained list of characteristics to be accounted for, this approach provides space for misusing CCV. Thus, one conclusion is that appraisers have certain discretion in selecting the peer group when applying the practice-oriented approach, which might lead to inaccuracies or even to manipulations of the valuation results.

Similarity criteria for multi-factor models

The similarity criteria for multi-factor models are usually lower than for single-factor models. This is the case because the aim of multi-factor models is to trans-

60 An in-depth analysis of the determinants of the long-term growth rate and the problems of predicting it can be found in sections 4.2.2.4 and 4.2.2.5.

fer certain similarity criteria directly into the valuation model. This is consistent with the overall principle of the development of CCV from immediate valuation models over single-factor valuation models to multi-factor valuation models: The more factors are explicitly part of the valuation model, the less similarity determinants are necessary. However, multi-factor models do not always succeed in totally eliminating similarity determinants. This becomes obvious when taking a closer view on the PEG ratio that at best only provides a small improvement – if any at all – compared to single-factor models.[61]

From the standpoint of comparable company selection the message is clear: It is desirable to develop a CCV model that has few and – if possible – directly observable comparable company selection criteria. In the best case, principally all companies can serve as a comparable company with the consequence that no similarity requirements are necessary. However, the development of powerful multi-factor models is not only determined by an uncomplicated comparable company selection process but also by other factors discussed later in this section.

3.1.2 Degree of Market Efficiency and Pricing Quality

3.1.2.1 Financial Theory

The existence of a certain degree of *capital market efficiency* and pricing quality is a necessary pre-requisite for accurately performing CCV. There are several forms of market efficiency, including allocational, operational and informational market efficiency (see Copeland et al., 2005: 353-354). For CCV purposes the informational efficiency is of particular importance. A market is informationally efficient if prices always fully reflect all available information (see Fama, 1976: 133; Fama, 1970: 383; Jones, 1998: 255). More precisely, "a market is efficient with respect to information set [U_t; the author] if it is impossible to make economic profits by trading on the basis of information set [U_t; the author]" (Jensen, 1978: 96). Consequently, the degree of informational efficiency crucially determines how the information about future financial benefits of comparable companies flows into their market prices and, thus, finally influences the result of the CCV process. Below, the term market efficiency purely refers to informational market efficiency.

Pricing quality is a more general term and refers to the degree to which stocks are mispriced. A stock is mispriced if its price deviates from its fair value. Market efficiency and pricing quality are highly related terms but, in fact, they are not identical. Market efficiency is always related to the information available. Thus, in an efficient market abnormal returns – and therefore cases of mispricing – are not possible on a systematic basis (see Gerke and Bank, 2003: 52, 93). However, the existence of an efficient market does not imply that there are no abnormal returns at all. Instead, it must be assumed that in an efficient market mispricing is possible, but its appearance is random and therefore not correlated to any value-

61 See section 4.3.1.1.

determining variable.[62] In an efficient market, the probability of a stock being underpriced is equal to the probability of a stock being overpriced. None the less, on average stocks are priced correctly.

The relationship between both terms is straightforward. Typically, a lower degree of market efficiency involves a lower or at least a less-predictable pricing quality. If a market is not perfectly efficient, then stocks are not on average correctly priced. Thus, in such a market it is possible to earn abnormal returns on a systematic basis. That also means that in a market that is not perfectly efficient the probability of overpricing does not necessarily equal the probability of underpricing. Therefore, the risk of the existence of generally overpriced or generally underpriced assets is higher in an inefficient market.

For purposes of CCV both market efficiency and pricing quality are important points. The following remarks will mainly focus on market efficiency since it is academically more intuitive to deal with available information. The academic determination of pricing quality of single stocks is principally possible, but a market-wide assessment would require the performance of a huge amount of individual accurate direct valuations and is therefore a nearly impossible task.[63] Additionally, it is not necessary here to exclusively focus on pricing quality since this section comprises the respective conclusions that can be drawn from the analysis of market efficiency for the pricing quality, and what these results mean in the context of CCV.

The theory of market efficiency is often regarded – along with the development of the CAPM and the Modigliani-Miller irrelevance proposition – as one of the three most important concepts of classical finance on asset pricing (see e.g. Gilson and Kraakman, 2003: 5-7). The central declaration of market efficiency theory, the efficient market hypothesis (EMH), states that there will be an absence of arbitrage opportunities in markets populated by rational, profit maximising individuals.[64] This, in turn, means that there are no trading systems or investment strategies that are based on the information set which can yield excess returns for any secu-

---

62 "Contrary to popular view, market efficiency does not require that the market price be equal to true value at every point in time. All it requires is that errors in the market price be unbiased; prices can be greater than or less than true value, as long as these deviations are random." Damodaran (2002: 113). See also Fama (1976: 144); Damodaran (2001a: 950); Eidel (2000: 13).

63 A similar opinion is provided by Brav and Heaton (2003: 12-13). In chapter 2 it has been shown that principally powerful direct valuation models exist. However, the human factor in the application of these models often leads to inaccurate valuation results; see Gilson et al. (2000), Francis et al. (2000). It is also important to note that the determination of pricing quality requires an absolute, direct valuation model, not a relative pricing model, resp. arbitrage pricing model (the CCV models, typical derivative pricing models or the Arrow-Debreu approach are examples for the latter category). The pricing quality of single stocks is also often referred to as "valuation efficiency", see Gerke and Rapp (1994: 10-12).

64 See Fama (1976: 134-137); an interesting study about whether arbitrage is powerful enough to keep prices efficient can be found in De Long et al. (1990a).

rity j. This can also be put formally with regard to the available information set U_t:[65]

$$E\left[EMV_{j,t+1} \middle| U_t\right] = 0, \quad a.s.$$
$$for \quad EMV_{j,t+1} = PRICE_{j,t+1} - E\left[PRICE_{j,t+1} \middle| U_t\right] \tag{3.4}$$

with $E\left[PRICE_{j,t+1} \middle| U_t\right]$ = the conditional expectation of the price of the security j at time t+1, given the information set U_t.

$EMV_{j,t+1}$ = the excess market value of security j at time t+1.

Note: The correctness of Equation 3.4 can be seen using (#) the Linearity Condition and (*) the Tower Property together with the fact that U_t is (the biggest) sub-σ-algebra of U_t itself.[66]

$$\begin{aligned} E\left[EMV_{j,t+1} \middle| U_t\right] &= E\left[PRICE_{j,t+1} - E\left[PRICE_{j,t+1} \middle| U_t\right] \middle| U_t\right] \overset{(\#)}{=} \\ &= E\left[PRICE_{j,t+1} \middle| U_t\right] - E\left[E\left[PRICE_{j,t+1} \middle| U_t\right] \middle| U_t\right] \overset{(*)}{=} \\ &\overset{a.s.}{=} E\left[PRICE_{j,t+1} \middle| U_t\right] - E\left[PRICE_{j,t+1} \middle| U_t\right] = \\ &= 0 \text{ , } a.s. \end{aligned} \tag{3.5}$$

For purposes of CCV this means that market efficiency is a necessary prerequisite to make the three arbitrage relations (Equations 2.7, 2.8 and 2.9) hold. To put it differently, even if appraisers manage to synthetically create similar assets, it still takes an efficient market for the arbitrage relations to be valid. However, the absence of an efficient market does not necessarily make CCV a defective valuation approach. CCV can still deliver important information in a rather inefficient market. Nevertheless, it is important to first assess the magnitude of the deviation from perfect market efficiency in order to judge the applicability of CCV since this amount of deviation crucially determines how far prices presumably differ from the intrinsic value of a company.

A closer look at the different forms that market efficiency can take on, particularly as from the perspective of behavioural finance towards market efficiency, and a comparison of how information is considered differently in CCV and in direct valuation approaches will shed more light on how to cope with market inefficiency in CCV.

---

65 The following is an extended version of the formal presentation of Fama (1970: 385).

66 For further information about conditional expectations, see Williams (2001: 83-92).

Different forms of market efficiency

The overall EMH is usually divided into three subhypotheses, which differ in the design of the information set. These three forms of the EMH are called the weak-form, the semistrong-form and the strong-form EMH.[67]

The *weak-form EMH* is based on the security market information set (also called the historical information set). This information set is composed of historical price data, historical rates of return, trading volume data and other historical market generated information. Historical corporate specific information is also part of the set. This form of the EMH assumes that current stock prices fully reflect all historical information about the respective securities. This especially implies that current returns should have no relationship with future returns; i.e. there is no serial correlation of security returns. For reasons of security analysis that means that no abnormal returns can be gained based on trading rules that rely on past price information. The weak-form EMH therefore calls the use of technical analysis into question.

The *semistrong-form EMH* is based on the public information set. Public information is all the information that is disclosed to the market place. This also includes the historical information set that is the basis of the weak-form approach. Furthermore, information about the future, such as, e.g., earnings estimates and material information concerning pending corporate transactions, is part of that information set. This form of the EMH asserts that stock prices rapidly adjust to the disclosure of new information and therefore at every point in time reflect all public information. It encompasses the weak-form EMH because the weak-form information set is a subset of the semistrong-form EMH. For investors the assumption of the semistrong-form EMH means that there can be no systematic gain from trading decisions based on newly released information. This, in turn, implies that fundamental analysis is without any use in security analysis.

The *strong-form EMH* is based on all information from public as well as from private sources. Due to this underlying ultimate information set, strong-form EMH encompasses the two other forms of EMH. It implies that there is no insider information since all information is available to everyone at the same time. Therefore, no investor should be able to yield systematic abnormal profits. From a theoretical point of view, this form of EMH has been excessively challenged by two closely related paradoxes. First, the no-trade theorem postulates that rational trading should not take place if nobody can expect to benefit from it (see Milgrom and Stokey, 1982). Second, every security analysis is redundant since market prices already include all information. If, however, nobody gathers information, then market prices cannot contain all information (see Gerke and Bank, 2003: 91). There is a solution to both paradoxes, which is straightforward and, thus, these paradoxes do not automatically lead to the rejection of this hypothesis, though: Some market participants simply do not believe in the strong-form EMH even if it actually

[67] The following descriptions are principally based on Reilly and Brown (2000: 215-216); Gerke and Bank (2003: 92-93); Copeland et al. (2005: 354-355); Elton et al. (2003: 402-404).

holds.[68] Typical strong-form EMH doubters are active money managers, arbitrageurs and brokers, whose businesses aim at finding market inefficiencies.[69] A positive consequence out of this is that their work helps to enhance capital market efficiency and therefore supports the strong-form hypothesis.

Market efficiency from the perspective of behavioural finance

The analysis of market efficiency from a behavioural finance perspective focuses on how arbitrage mechanisms work on real capital markets with rational and irrational market participants. This approach to market efficiency has found a lot of support because it might help to solve the market indeterminacy issue, which describes the problem of failing to determine whether markets are efficient due to a lack of reasonable pricing models.[70] The central statement of the behavioural finance approach is that the existence of market inefficiencies does not automatically lead to a valuable arbitrage opportunity because correction strategies can be risky and costly (see Barberis and Thaler, 2002: 4; Gilson and Kraakman, 2003: 17-18). More precisely, the existence of certain classical arbitrage risks makes it highly unlikely that perfect market efficiency exists.

These classical risks of arbitrage fall into three categories. (1) The fundamental risk describes the risk that not only the assets under consideration but also existing substitutes are mispriced. Consequently, the arbitrage process would fail because arbitrage prices are always linked to the comparable asset (see Barberis and Thaler, 2002: 5). (2) The noise trader risk denotes the risk that prices deviate systematically from fundamental values for longer time horizons because irrational investors continue trading the security while maintaining the level of irrationality or even becoming more irrational over time.[71] This would lead to the situation in which it is not possible for rational investors to benefit from arbitrage even if they detect systematically mispriced securities. (3) The implementation cost risk refers to the risk that the existence of process costs makes it less attractive for rational

---

68 Grossman and Stiglitz argue that the existence of these active doubters and of arbitrageurs in general is in turn a proof of existing market inefficiencies or – more precisely – of an "equilibrium of disequilibrium", see Grossman and Stiglitz (1980: 393). The author of this book agrees with that reasoning. However, minor deviations from the equilibrium do not call the general principle of the strong-form EMH into question. Thus, the following theoretical outlines are valid, even if perfect strong-form efficiency cannot exist.

69 This point has been put very colourfully by Rex Sinquefield, former chairman of the US investment advisory company Dimensional Fund Advisers: "There are three classes of people who don't think that markets work: the Cubans, the North Koreans and active money managers.", LeBaron et al. (1999: 6).

70 See Roll (1977); Fama (1991: 1576) for a description of the "joint-hypothesis" problem that might lead to market indeterminacy; for a more thorough discussion about market indeterminacy, see Brav and Heaton (2003).

71 See De Long et. al. (1990b), Shleifer and Vishny (1997). Black did not see a general problem for market efficiency due to noise. He (pragmatically) proposed in this context that markets should still be named efficient if prices are within the factor "2" of the intrinsic value, see Black (1986: 533).

investors to exploit existing arbitrage opportunities. These costs can be typical transaction costs, such as commissions, fees or the bid-ask spread. However, the cost arising from certain short-sale constraints, the costs of detecting any existing mispricing and the cost of actually exploiting that mispricing can also add to prevent arbitrage (see Merton, 1987).

Because of these classical arbitrage risks market participants sometimes remove from the typical proceeding of arbitrage, which aims to exploit differences between the fundamental value of a security and its price. Rather, they switch to a strategy of temporal arbitrage between the stock's current market price and the future expected market price, regardless of how the future expected market price and the fundamental value are related. To put it differently, the temporal arbitrage strategy aims to yield positive returns by anticipating the valuation error of irrational investors while considering the implementation cost risk. This behaviour, in turn, might lead to an aggravation of existing market inefficiencies (see Gilson and Kraakman, 2003: 24).

A comparison of the information processed in direct valuation approaches and CCV

For reasons of simplicity it is assumed below that there are a reasonable number of comparable companies that perfectly satisfy the comparability criteria. This assumption is important since as long as adequate comparable companies can be found, the information content of CCV equals the information content of stock prices. However, to better assess the amount and quality of the information embedded in stock prices, it is indispensable to compare it with the information processed in *direct valuation approaches*.

The information content of direct valuation approaches is typically difficult to assess across-the-board. It greatly depends on the experience, motivation and access to information of the appraiser.[72] Therefore, different states of information content have to be distinguished. In the extreme case of a fully informed appraiser, the result of a direct valuation process combines insider knowledge about the company with broad public information. The information processed in such a valuation case equals that of prices in perfectly efficient capital markets. However, it is unlikely that such a valuation case is of any major practical relevance for two reasons: First, the number of people with access to non-public information is typically small, so that only a small percentage of all valuations incorporate material private information. Second, even if appraisers possess material private information, to cope with ethical requirements they are advised neither to trade on it nor use it for a report. Instead, they should try to get the private information transmitting person to disseminate that information to the market place (see AIMR, 1999: 236).

72 For reasons of simplicity the following outlines exclusively focus on the information content. This especially means that it is assumed that prices and values do not differ in other characteristics (in the sense of the discussion in section 2.2.2) than the information content.

Loosening the assumptions a bit leads to the situation in which an appraiser has access to all public but no private information. In such a case, the information content of a direct business valuation would equal the prices in semistrong-form efficient markets. Although theoretically possible, doubts have to be entertained as to whether appraisers are able to collect all public information about a stock. Such a collection would require an immense period of time and funds and, thus, make the valuation process a very costly task that additionally lacks timeliness. Therefore, in a more realistic scenario, the typical average appraiser has no private but certain public information. However, he does not posses all public information.

The amount of information processed in capital markets, in turn, crucially depends on the information input of market participants. Capital markets, however, are able to put together information of single market participants and therefore process a larger amount of information. This ability is also known as "aggregation efficiency" of capital markets and is a subset of information efficiency. An aggregation efficient market is a system that puts together partial information to create a signal – the market price – that incorporates the whole information collected. As a consequence, all market participants eventually have access to the collectivity of information. This kind of efficiency is the result of the competition of market participants over the best investments (see Rubinstein, 1974; Gerke and Rapp, 1994: 11-12; see Weber and Wüstemann, 2004: 6-8 for some practitioners' support for this thesis).

The existence of aggregation efficiency leads to the conclusion that capital markets usually process a higher amount of information than each single investor – or as Brav and Heaton put it: "We have no doubt that market prices may incorporate far more information than any person may possess."[73] Thus, from a theoretical standpoint it can be stated that prices at capital markets should be expected to carry a superior amount of information than direct company valuations do –

[73] Brav and Heaton (2003: 13). Copeland et al. also conclude that the information content of stock prices is very high in general: "Market efficiency implies that stock prices reflect all available information. We recommend using this information as much as possible in corporate decisions." Copeland et al. (2005: 378). An interesting view on the relationship between the information of single investors and that of the market has been provided by Roll (1984). He examined the relationship between the official weather forecast for the region around Orlando, Florida, and the pricing on the US orange juice futures market (at the time of the study 90 % of the oranges used for frozen orange juice concentrate grew in that region; Roll comments that it is widely recognised that the weather is the variable with the most important influence in this market). One important result of his analysis was that, if one uses the pricing information to forecast the weather, the forecasts are far better than those made by the weather bureaus. Thus, the market did not only incorporate the whole information provided by the weather bureaus, it also had more information than any single forecast provider had.

even if the speed of information aggregation in prices is not independent of the respective amount of information that underlies the valuations of single investors.[74]

It is important here to get back to the initial assumption of this sub-section. In fact, these comparisons of prices and direct valuation results can only be carried forward to CCV if perfect substitutes are available for the target company. If, however, the comparable companies are not perfectly similar, then parts of information that are target specific cannot be considered in the CCV process. This, in turn, lowers the amount of information incorporated in CCV.

3.1.2.2 Factual Market Efficiency

International empirical evidence

Two categories of tests are typically used to examine the existence of *weak-form efficieny*. The first approach focuses on the prediction of the weak-form EMH that stock returns should not exhibit dependencies over time. Serial correlation tests over short time periods largely support the weak-form EMH[75], even if weak autocorrelation exists for certain portfolios of stocks (see Conrad and Kaul, 1988; Lo and MacKinlay, 1988). Mean reversion of returns is, however, observable over long and short time horizons (see DeBondt and Thaler, 1985; Oehler, 1995; for Germany see Meyer, 1994 and Schiereck and Weber, 1995). Runs tests, which compare the actual number of consecutive price changes into one direction with the expected number of these price changes, also typically fail to reject the null-hypothesis of weak-form EMH (see Hagermann and Richmond, 1973; Fama, 1965; Jennergren and Korsvold, 1975; Elton et al., 2003: 413). To better account for the complexity of certain technical analysis tools which go far beyond a set of simple price changes, weak-form efficiency is also tested by a second approach: The simulation of specific trading rules (see Reilly and Brown, 2000: 217-218). The results of these tests are mixed: Some do not find that one can systematically yield abnormal returns with these strategies (see Ball et al., 1995) while others support the theory of success of specific trading rules (see Pinches, 1970; Brush, 1986; Pruitt and White, 1988; Lakonishok and Vermaelen, 1990; Mitchell and Stafford, 1997). Recent research, however, shows that after accounting for commissions, most trading strategies fail to beat a simple buy-and-hold policy (see Bessembinder and Chan, 1998). This all leads to the conclusion that capital markets in general seem to be weak-form efficient.[76]

---

[74] See Gilson and Kraakman (2003: 10). Sometimes it is pointed out that stock prices incorporate a higher rationality than individual valuations because stock prices manage to even out irrational moods of single investors, see Cochrane (1991: 483).

[75] See Fama (1965); Fama and McBeth (1973); Elton et al. (2003: 410-411); Ross et al. (2005: 359); over longer time periods negative correlations of down to -0.4 have been found, see Poterba and Summers (1988); Fama and French (1988). However, since 1940 these correlations have not beeen significant.

[76] This can also be assumed for the German Market, see Loistl (1994: 172); Möller (1985: 514-516). A contrary opinion is provided by Bräutigam (2004) who found non-linear timely patterns for the German stock market.

Tests of *semistrong-form efficiency* can also be divided into two categories: return prediction studies using public information on the one hand and event studies on the other hand (see Reilly and Brown, 2000: 219). *Time-series tests* as part of the return prediction tests aim to determine whether any public information can provide superior estimates of future returns than historical long-run returns. While these tests provide limited support for a significant impact of public information in the short-run, they are quite successful in showing its influence in the long-run: Several analyses demonstrate certain, even though mostly low, stock returns predicting ability of the dividend-yield alone (see Rozeff, 1984), of the dividend-yield combined with variables describing the term structure of interest rates (see Keim and Stambaugh, 1986; Chen, 1991) and of a number of business-cycle variables (see Pesaran and Timmermann, 1995). These results question the existence of semistrong-form market efficiency. Moreover, studies that examine the usefulness of disclosed earnings in predicting future asset prices do not provide much support for the semistrong-form EMH, either (see e.g. Watts, 1978; Foster et al., 1984).

Cross-sectional studies reveal that return anomalies exist for certain groups of companies, such as the overperformance of low PE ratio firms compared to high PE ratio firms (see Basu, 1977; Peavy and Goodman, 1983; for Germany see Garz, 2000: 137-139), the positive relationship between the PB ratio and future stock returns (see Fama and French, 1992; Dennis et al., 1995; Fama and French, 1998; for Germany see Garz, 2000: 134-135), the larger risk-adjusted returns for small firms compared to large firms (size effect) (see Banz, 1981; Reinganum, 1981; Reinganum, 1992; for Germany see Beiker, 1993: 185-458; Garz, 2000: 130-133) and the return deviations for stocks that are thinly traded or that are covered by only a few analysts (neglected firm effect) (see Arbel and Strebel, 1983; James and Edmister, 1983). Additionally, certain so-called calendar effects have been found which indicate that timely regularities exist in stock returns. Examples of these effects are the January-effect[77], which stands for the significantly higher returns in January compared to the rest of the year, and the "day of the week"-effect[78], which highlights the findings that mean returns for Monday are significantly negative.

Since all of these analyses are necessarily joint tests of both the market efficiency and the applied asset pricing model and therefore anomalies can be the result of existing inefficiencies or a mis-specified pricing model[79], no definite statement can be made regarding the consequences for the semistrong-form EMH. It has to be assumed, however, that the anomalies are at least partially due to market

---

77 See Roll (1983); Reinganum (1983); Tinic and West (1984) and Riepe (2001); weak evidence for Germany is provided by Stehle (1991) and Wallmeier (1997).

78 See French (1980); Keim and Stambaugh (1984); Athanassakos and Robinson (1994); international evidence is provided by Dubois and Louvet (1996); for Germany see Frantzmann (1987).

79 See Ross (2005: 49-50); see Roll (1977) for a critical assessment about the use of the CAPM as a reference model ("Roll's Critique"). As regards other problems with tests of semi-strong form efficiency, see Copeland et al. (2005: 389-390).

inefficiencies. Nevertheless, during the last few years some of the cross-sectional anomalies seem to have disappeared or have at least lost significance.[80]

Event studies as the second approach to semistrong-form EMH testing have enjoyed dramatically increasing popularity in practice during the last decades. The goal of these studies is to analyse whether it is possible to derive abnormal rates of return during a certain time period immediately after an announcement of a significant economic event. Contrary to what the time-series and cross-sectional tests revealed, the results of most of the event studies support the semistrong-form EMH. More precisely, markets typically react more or less quickly and in a reasonable manner to the announcement of stock splits, additional listings at or changes to a major stock exchange, political or economic world events, corporate events and announcements about certain corporate accounting changes.[81] Additionally, markets usually quickly correct an existing underpricing of IPOs[82]. In conclusion, evidence from tests of the semistrong-form EMH is mixed: While the results of event studies usually support this form of efficiency, several time series and cross sectional analyses cast doubt on it.

Tests of the *strong-form EMH* usually focus on the performance of certain groups of investors who are supposed to have private information. Supporters of this variant of the EMH should expect none of these investors to consistently earn abnormal profits. Following the results of the studies, professional money managers – who are supposed to be the best-informed investors without monopolistic access to new information (see Reilly and Brown, 2000: 244-245) – are not able to systematically beat a buy-and-hold investment strategy (see Sharpe, 1966; Jensen, 1968; Shukla and Trzcinka, 1994; Rubinstein, 2001). Analyses of financial analysts' performance deliver mixed results: Some examinations show analysts' ability to beat the market (see e.g. Stickel, 1985; Womack, 1996); others cannot find evidence for a superior performance (see e.g. Desai and Jain, 1995). Groups that have monopolistic access to insider information, like corporate insiders and stock exchange specialists, can usually derive abnormal profits (see Finnerty, 1976; Givoly and Palmon, 1985; Seyhun, 1992; Pettit and Venkatesh, 1995; for Germany see Seeger, 1998: 269-274). However, non-insiders cannot benefit from the

80 See e.g. Kothari et al. (1995) regarding the PB effect, Fortune (1991) regarding the size effect and the January effect, Beard and Sias (1997) regarding the neglected firm effect and Wang et al. (1997) regarding the day-of-the-week effect.

81 For an overview of event studies covering these influences, see Reilly and Brown (2000: 234-239). A study for the German and the Austrian stock market, however, found that these reactions sometimes take several days, see Holzer (2001: 126); Gerke et al. (1999) showed that market participants can derive abnormal returns (even after the date of the announcement) from trading stocks that are newly included into certain German stock market indexes, which is also not consistent with semistrong-form efficiency. Contrary to that, Nowak (2000) concluded based on the results of studies about the price impact of accounting information in Germany that the German market has a sufficiently high degree of information efficiency, see Nowak (2000: 191).

82 For an overview of US studies with respect to IPO underpricing, see Copeland et al. (2005: 391-394).

knowledge of superior performance of these latter insiders (see Lee and Solt, 1986). To summarise: Strong-form efficiency does not exist in general. This is the case because insiders with monopolistic access to private information can derive abnormal profits. However, other market participants, seen as insiders without monopolistic access to non-public information, have not been able to beat the market.

The behavioural finance approach to market efficiency has also recently been tested. One study provides weak evidence for the existence of noise trader risk (see Froot and Dabora, 1999); another one shows that implementation costs can severely limit arbitrage opportunities in corporate transactions like carve-outs (see Lamont and Thaler, 2003). Nevertheless, the bottom line is that there is no broad empirical evidence for the existence of arbitrage risks in general.

The development of the German stock market during the last few years

A central problem all empirical studies about market efficiency have is that most of them were conducted several years ago. For market participants, however, it is rather interesting to know how efficient capital markets are today. Some assistance in answering that question might come from a short overview on how capital markets have experienced structural changes during the last years. Most of these changes were implemented in order to improve market efficiency and it can be assumed that they eventually succeeded in doing so. The German market, for example, has made noticeable efforts in regards to the trading volume of single stocks, the number of listed companies, the growing share of international investors and the internationalisation of accounting rules (see Löhnert and Böckmann, 2005: 407-408; Solnik and McLeavy, 2004: 141). Additionally, new regulations like the Law for the Control and Transparency in the area of organisations (1998, Gesetz zur Kontrolle und Transparenz im Unternehmensbereich, KonTraG), the Capital Raising Promotion Act (1998, Kapitalaufnahmeerleichterungsgesetz, KapAEG), the German Corporate Governance Code (2002) in conjunction with the Transparency and Public Disclosure Act (2002, Gesetz zur Transparenz und Publizität im Unternehmensbereich, TransPuG)[83], the Investment Modernisation Act (2003, Investmentmodernisierungsgesetz, InvModG) and the Law on the Improvement of Investor Protection (2004, Anlegerschutzverbesserungsgesetz, AnSVG) – to name but the most important – have clearly improved the information flow from listed companies to investors and added to an enhanced attractiveness to the German stock market.

What analysts and investors think about German stock market efficiency in 2005

In order to draw a complete picture of efficiency and to get a comprehension of the current state of market efficiency, 222 financial analysts and institutional investors were asked in April 2005 about their assessment of informational efficiency on the German stock market. This survey was conducted as part of the

---

83 See Meitner (2003) for an overview of how the German Corporate Governance Code has influenced the transparency of listed companies.

monthly financial market survey of the Centre for European Economic Research (ZEW). One important benefit from this survey is that it might provide a way to overcome (or at least to mitigate) the market indeterminacy problem since financial analysts do not exclusively rely on statistical valuation models but typically base their overall impression about stock market valuation on different sources of information and different valuation approaches.

The results from this survey largely support the outlines above. According to the analysts, informational efficiency at the German stock market has well developed during the last years. There is considerable improvement as compared to the period before 1990 and still moderate improvement as compared to the early 1990s and to the time of the dot.com boom. Figure 6 illustrates this (a value of 4 indicates indifference).

Fig. 6. Development of informational efficiency on the German stock market (in the eyes of analysts and investors)

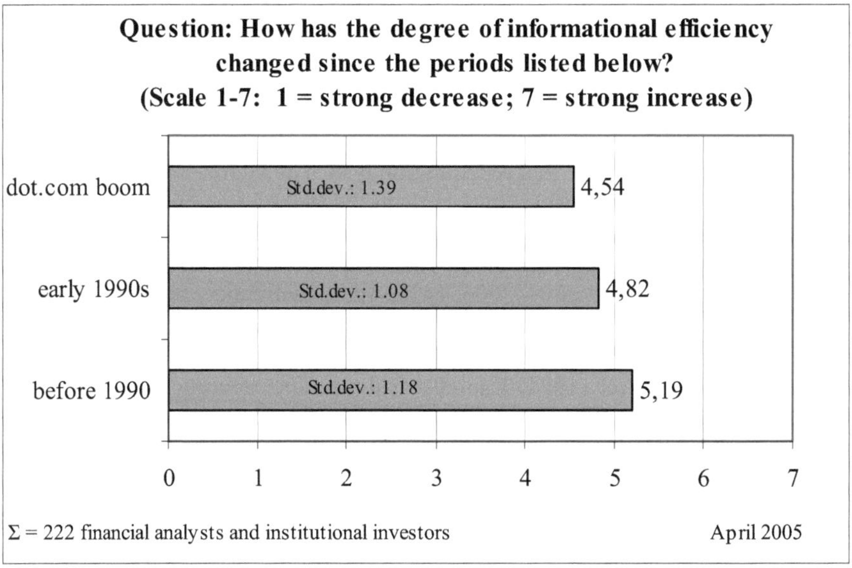

Source: ZEW

Asked about the main reasons for that improvement, 56.2 % named the internationalisation and growing experience of market participants, while 53.9 % suggested that the regulations and laws that aim at enhancing capital market transparency drove market efficiency (more than one answer was possible). The increasing use of international accounting rules (38.8 %) and developments in trading systems (36.5 %) also have some impact. The voluntary Corporate Governance Code was given inferior relevance (30.3 %).

Additionally, more than 75 % of the survey participants believe that there is no mispricing or only some unsystematic mispricing in the German stock market index DAX (see figure 7). This finding also strongly supports the semistrong EMH.

Fig. 7. The difference of market prices and intrinsic values of DAX-companies (in the eyes of analysts and investors)

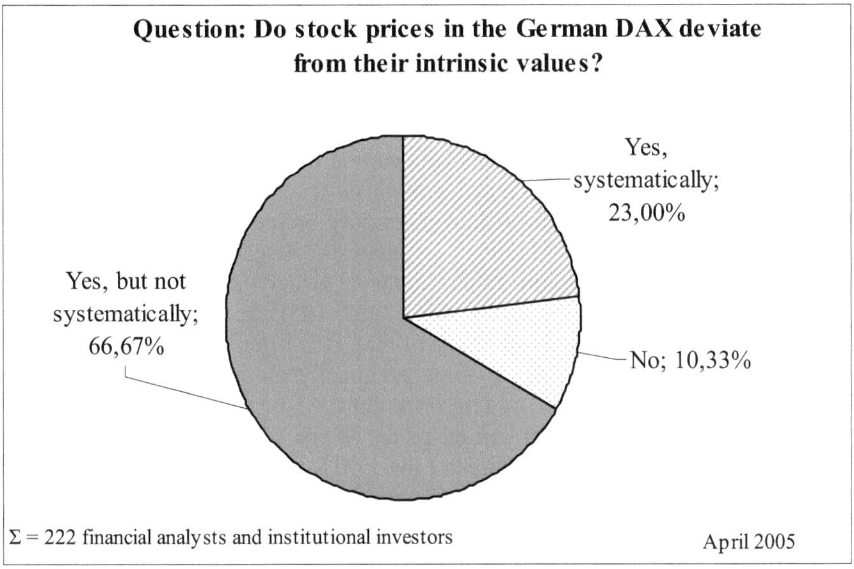

Source: ZEW

As the variety of empirical evidence outlined above may suggest, it is difficult to give an overall assessment of market efficiency. At the current state of research, the bottom line is that capital markets may not be perfectly efficient in an academic sense, but they are of high efficiency in general.[84] The weak-form EMH cannot be rejected by empirical research and the semistrong-form EMH also enjoys widespread support. However, there are anomalies, which give rise to the supposition that certain violations of the semistrong-form EMH exist, even if they may also be due to a mis-specified pricing model. Nevertheless, none of these anomalies has enough impact to principally call that form of market efficiency into question. However, there is strong evidence that strong-form efficiency does not exist: Even if experts cannot outperform the market, monopolistic insiders can systematically earn abnormal profits.

Summary about market efficiency

The recent developments in the legal framework of German capital markets, however, give reason to the supposition that capital market efficiency has increased

[84] See also Brealy and Myers (2003: 370); Solnik and McLeavy (2004: 141). Ross (2005) suggested based on hedge fund activities that inefficiencies only arise for about 0.1 % of all assets in world markets. He concluded: "There should be no debate, then, over whether the efficiency glass is half full or half empty; it is simply quite full", Ross (2005: 64).

during the last few years. This is consistent with what financial analysts and institutional investors think about the German market.

3.1.2.3 A Short Note on Crashes and Bubbles

The issue whether crashes or bubbles[85] are indicators of market inefficiency is not addressed in-depth here. Both phenomena can be explained by a rational change in, e.g., investors' expectations about growth rates of dividends (see Brealy and Myers, 2003: 362) or investors' risk aversion (see Steiner and Bruns, 2002: 296). The fact that most DCF valuations in analysts' research reports during the dot.com-boom yielded corporate values that were largely identical to the then prevailing market prices (see Beckmann et al., 2003: 105) clearly supports the theory of rational investor behaviour. Contrary to that, it is often argued that these phenomena are driven by irrational, psychodynamically motivated investors' behaviour such as positive feedback trading or herding.[86] Consequently, no clear statement with respect to the EMH can be made: While the theory of rationality in crashes and bubbles does not contradict the EMH, the theory of psychodynamic influence (that is based on behavioural finance) does.

The central problem here is that because of market indeterminacy it is not possible to assess whether prices really move away systematically from intrinsic values in these situations. For reasons of practicability, it is assumed that even at times of crashes and bubbles there are no major differences between the information processed in DCF valuation and that inherent in market prices. This seems to be a plausible assumption since – even if there is irrationality in asset prices – common valuation models typically cannot cope with that irrationality (see Bruns, 1994: 210-212).

3.1.3 Consequences for Comparable Company Valuation with Special Regard to the Peer Group Selection

A few general remarks regarding the comparable company selection process can be made depending on the availability of companies, the degree of market efficiency, the relationship between the information processed in direct valuation approaches and the information incorporated in market prices, the applied selection approach and the valuation model used. All statements are based on the assumption that there are no factors other than the degree of similarity and the market efficiency that might influence the selection process:

---

85 Typically a bubble is understood as the difference between intrinsic value and price; see Gruber (1988: 20). However, here it is understood (more generally) as a time of strongly rising stock prices independent of the relationship to intrinsic values.

86 See Bruns (1994: 93-108), if so then the general validity of the weak form EMH has to be called into question, see Oehler (1995: 116).

a. The availability of similar companies crucially determines the necessary number of peer group companies. In general, a high number of comparable companies is desirable if companies are not perfectly similar and differences in certain criteria are normally distributed around the target company's criterion value. Thus, with a higher number of comparable companies, appraisers typically manage to even out these differences.[87] It can also be stated that the higher the similarity of companies, the fewer peer group companies are needed. Assuming a highly efficient relevant market with no mispricing, it is even sufficient to apply only one comparable company in case this comparable company perfectly matches with the target company.
b. The big problems and expenses that are attached to the determination of the sustainable growth rate for all peer group companies as well as the target company gives rise to the fact that the pure valuation theory-driven approach does not enjoy widespread popularity in practice. Appraisers therefore usually rely on the practice-oriented approach, a fact that, however, does not mean that the selection process is an in-and-out process. In any case, the time input for investigations and analyses necessary here is absolutely comparable to that of direct valuation approaches (see Moser and Auge-Dickhut, 2003: 222).
   The use of the practice-oriented approach rather than the valuation theory-driven approach is usually done at the cost of the number of comparable companies (if properly applied). Since the practice-oriented approach requires a larger amount of similarity factors it is less probable to find a similar company when using that approach. Additionally, appraisers face high uncertainty as to whether the chosen characteristics can really substitute the growth rate. The need for balancing that expected lower probability of similarity and the uncertainty about the similarity lead to a relatively higher required number of comparable companies.
c. If there is no adequate similarity in relevant criteria, then companies have to be selected to have *on average* adequate similarity in reference to these criteria. This especially means that it is not reasonable to only use comparable companies that are biased in one direction compared to the target company for certain similarity criteria. Problems with that issue typically arise if the target company is basically different from all companies principally available for comparing (e.g. the target is market-leader in its segment or a segment underperformer).
d. If the target and the companies that can principally serve as comparables differ in certain qualitative aspects that cannot be averaged (i.e. the legal environment, capital market orientation) then only those companies should be included into the peer group that are identical in the relevant criterion.
e. The number of comparable companies required has to be regarded against the background of the limited number of existing exchange listed companies. The limited number of existing companies obviously also limits the number of the basic population for the selection process (i.e. companies that could principally

---

87 See letter g) in this enumeration for how a higher number of comparable companies can add to improve valuation accuracy assuming a normal distribution of the differences between peer group and target company.

serve as a comparable company). If, in turn, the number of comparable companies required is higher than the basic population, a lack of adequate comparable companies might be the consequence. Such a lack of comparable companies can lead to inaccurate valuation results and even to the total failure of CCV.

f. The degree of informational efficiency is obviously quite high at major international stock markets. This clearly indicates that the market approach to CCV is basically a reasonable valuation variant. However, empirical research showed that there are several semistrong-form anomalies that seem to be at least partially due to market inefficiencies. Most of these anomalies do not refer to a general mispricing of stocks but rather to a slower than expected reflection of current information in market prices. In order to avoid risiking a general bias of aggregated prices of peer group companies, appraisers are advised to exclude companies from the peer group that are affected by any of these semistrong-form anomalies (e.g. it might be better to remove small companies from a set of big companies; it might be better to remove firms not covered by analysts from a set of covered companies).[88]
However, with respect to the rejection of the strong-form EMH and the observed strong price movements in crashes and bubbles, a general statement about how advantageous the relative valuation approaches are as compared to direct valuation approaches is necessary: From the standpoint of information processing, CCV is only a reasonable valuation approach if the same amount of information or even more information is incorporated in market prices than could be incorporated in direct valuation approaches. As has been outlined in section 3.1.2.1, many real valuation settings – even situations like crashes and bubbles – can probably be characterised by this information distribution. This especially means that CCV can be a superior valuation approach compared to direct valuations, even if markets are not perfectly efficient. Contrary to that, if appraisers have access to all information necessary to perform a reasonable direct valuation and in case of justified doubts about the informational quality of market prices, CCV is not an appropriate valuation alternative. The limits of the advantageousness of one of the two valuation approaches are, however, blurred. In most cases it is incumbent upon the appraiser to subjectively decide. Certainly, the amount of information incorporated is not the only factor in determining the power of a valuation approach, but it is an important one. Nevertheless, the remarks above do not constitute a general judgement about the favourability of different valuation approaches.

g. Since it is impossible to determine whether stocks in general are correctly priced, appraisers have to rely on an assessment of the informational efficiency of markets. The relationship between informational efficiency and pricing quality (valuation efficiency) is such that a certain degree of informational efficiency implies that stocks are on average correctly priced based on the relevant set of information. This means that market participants use on average the given information correctly but it does not mean that market participants use it

[88] For an interesting discussion about the relevance of market prices in business valuation, see Weber and Wüstemann (2004: 22-24).

correctly for every single stock. Regarding the comparable companies selection process, this relationship implies that a higher number of comparable companies is always preferable over a lower number of comparable companies. This becomes clearer when looking at a set of stock prices that are standard normally distributed around the fair value.[89] The probability that a single stock is correctly priced is zero – i.e. $p\left(PRICE_j = E\left[PRICE_j\right]\right) = 0$.[90] Contrarily, the probability that the average of a high number of stocks (measured as the arithmetical mean) equals the fair value is – at the limit – 100 per cent – i.e. $\lim\limits_{M\to\infty} \sum\limits_{m=1}^{M} PRICE_m \cdot M^{-1}$ converges to $E\left[PRICE_m\right]$ a.s. (see Hogg and Craig, 1995: 240). This eventually implies that the probability that a fair value substitute can be synthetically created increases with the number of comparable companies.

h. It is worth noting that in real valuation settings, the number of comparable companies to be included in the valuation approach is limited by the number of comparable companies available with regard to the similarity criteria. When determining the number of comparable companies, appraisers have to consider that the improvement of the pricing accuracy due to a higher number of companies usually comes at the cost of a decrease of the peer group quality due to potential imperfect substitutes. Similarly, appraisers that choose a lower number of comparable companies usually face fewer peer group quality risks, but the risk of pricing inaccuracies is higher. This dilemma is a major problem in the process of selecting comparable companies

3.1.4 Implications for the Choice of the Valuation Model

The discussions about the determinants of the peer group selection process have a direct impact on the other main part of CCV: the choice of the valuation model. Two points are especially worth some remarks. (1) Sometimes appraisers do not have access to the relevant information to decide whether companies qualify as comparable companies and (2) sometimes there is simply a lack of companies that could principally serve as comparable companies (see Buchner and Englert, 1994: 1580, Löhnert and Böckmann, 2005: 406). As a consequence, from an appraiser's point of view it might be desirable in special valuation situations to apply valuation models that only have low similarity requirements. This would be a solution to both problems mentioned above, since (1) the lower the requirements are, the

---

89 The assumption of a standard normal distribution is made for illustration reasons. The results, however, also hold qualitatively for every other equal distribution around the fair value.

90 Even if small deviations from the fair value are tolerated, the probability does not increase much. E.g. the probability that the observation falls between ± 0.1 standard deviations of the fair value is still only about eight %.

less information has to be collected, and (2) the lower the requirements are, the more companies can principally serve as comparable companies.

This gives rise to the suggestion that – in case appraisers face a lack of information or a low number of principally comparable companies – one idea could be to switch from single-factor models to multi-factor models. As discussed in section 2.3.3, multi-factor models aim to decrease the peer group selection requirements while contemporaneously considering several value driving variables directly in the valuation model. However, whether certain multi-factor models really succeed in loosening the similarity requirements has not yet been extensively analysed. Therefore, this issue is in-depth addressed in section 4.3.

Fig. 8. Impact of similarity and pricing quality on the choice of the valuation models and the number of peer group companies needed

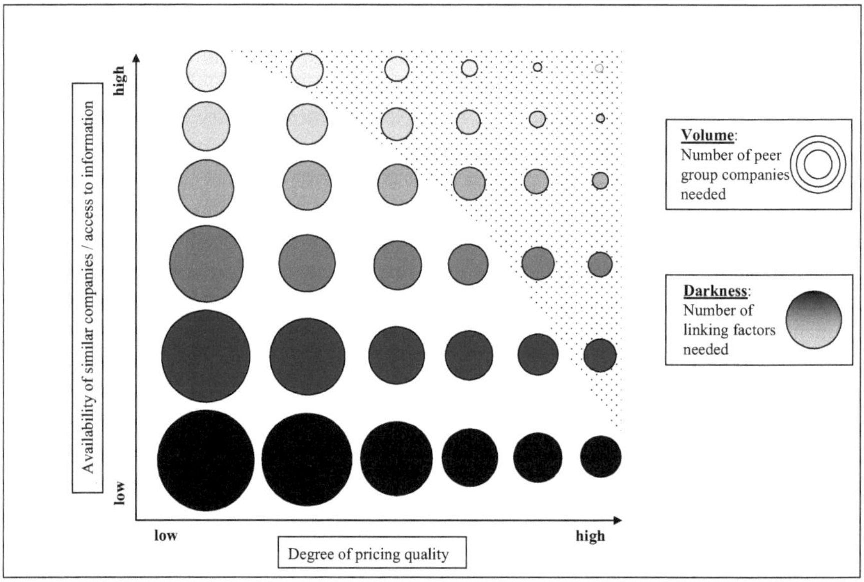

Source: ZEW

The need for multi-factor models in valuation cases characterised by bad access to information and lack of comparable companies is highlighted in Figure 8. This figure also emphasizes that the number of peer group companies needed is a function of the information access and the availability of similar companies as well as the pricing quality of the relevant market, too. The dotted area comprises the situations in which CCV can reasonably be performed. Outside of this area, CCV would typically fail in valuation practice.

3.2 Determinants of Valuation Model Choice

The valuation model choice is mainly driven by three external determinants. The first one is the value relevance of the reference variable (for single-factor models) or reference variables (for multi-factor models). This factor describes the relationship between the value of the target company and the valuation model, as well as the explanatory power of the valuation model. The second determinant is the future similarity between the target company and the peer group companies. Contrary to value relevance, this issue does not address the relationship between the target company and the valuation model but rather addresses the relationship between the comparable companies and the valuation model. Finally, the third determinant is the existence of technical limitations of valuation models. Obviously, a model can only be applied properly if there are no technical limitations implied by a certain valuation setting.

3.2.1 Value Relevance of the Reference Variables

3.2.1.1 Relevance of Accounting Variables for Valuation Purposes

The idea of analysing the relationship between stock prices (resp. corporate values) and accounting figures is not new at all. Researchers have devoted considerable effort to this topic since the late 1960s.[91] The reason for this widespread popularity is that all capital market participants (investors, issuers, researchers) strive for explanations of stock prices, and financial statements are usually the primary source of corporate information.

During the last decades three branches of this stream of accounting research have emerged (see Lo and Lys, 2000b: 3-14): information content studies, valuation relevance studies and value relevance studies. *Information content* studies concentrate on whether the announcement of new accounting variables leads to price changes that are in excess of the expected returns. Assuming the market efficiency framework provided in section 3.1.2.1, accounting variables in general have information content if and only if $PRICE_{it} \neq E\left[PRICE_{it} \,|U_{t-1}\right]$ with $PRICE_{it}$ being the (efficient) stock price incorporating the information set U_t, U_t being the set of information just after the announcement and U_{t-1} being the set of information just before the disclosure.[92]

From an accounting perspective, studies dealing with *valuation relevance* must be seen as a more specific approach to the relationship of accounting figures and stock prices. These studies focus on specific accounting items rather than on a

91 The seminal works of Beaver (1968) and Ball and Brown (1968) are widely seen as the starting point of this stream of accounting research.

92 The information content studies are based on the examinations of Beaver (1968). More recent studies are conducted, e.g., by Kim and Verrecchia (1991) and Bamber et al. (2000).

loosely specified set of information. However, the proceeding is similar to that of the information content studies. Valuation relevance exists if the unexpected change in stock prices can be consistently explained by the unexpected change in the accounting figure under examination. To put it in more formal terms:[93] an accounting figure *ACC* is valuation relevant if and only if the function *f (.)* in the following equation is non-trivial[94]:

$$\left|PRICE_{jt} - E\left[PRICE_{it} \mid U_{t-1}\right]\right| = f\left|ACC_{it} - E\left[ACC_{it} \mid U_{t-1}\right]\right| + \mu_{it}$$

with $\mu \sim \mathrm{N}(0,\sigma)$.

In valuation relevance studies the use of earnings as the accounting figure enjoys widespread popularity. In this context, the slope coefficient of a linear regression of unexpected returns on unexpected earnings is called the "earnings response coefficient" (ERC). The general conclusion is: The higher the ERC, the higher the valuation relevance and quality of earnings is (see Collins and Kothari, 1989; Ramesh and Thiagarajan, 1995).

Value relevance, finally, focuses on the direct relationship between market values and accounting variables. Contrary to the first two approaches described above, no expectation values of either future returns (as in both previous approaches) or accounting numbers (as in the valuation relevance approach) are needed.[95] The central question of value relevance studies is: Do certain accounting variables explain current market values? Or to put it more formally:[96] An accounting figure *ACC* is value relevant if the function *g(.)* in the following equation is non-trivial:

$$PRICE_{it} = g\left(ACC_{it}\right) + \eta_{it} \quad \text{with } \eta \sim N(0,\sigma).$$

If the analysis deals with more than one accounting figure, and, thus, the question is whether these accounting figures together exhibit some value relevance, then the equation that has to fulfil the non-triviality condition changes to:

$$PRICE_{it} = g\left(ACC_{1it}, ACC_{2it}, ACC_{3it}, \ldots\right) + \eta_{it}.$$

Facing these three different approaches, a question remains: Which of the three relevance-approaches is relevant for CCV? To answer this question, it is necessary to go back to Equation 2.4 (section 2.2.3.3). This equation is the formal description of the general CCV model. Equalling value and price for reasons of simplicity and assuming that the right hand side of the function is only a noisy estimate of

93 See Lo and Lys (2000b: 5-6). The valuation relevance studies are based on the examinations of Ball and Brown (1968).

94 Although there is no formal definition of trivial and non-trivial functions, a function is typically called non-trivial if the function value directly depends on the argument. A typical trivial function is, e.g., $f(x)=c$.

95 For a formal distinction between these three branches of capital market accounting research, see Lo and Lys (2000b: 7-12).

96 See Lo and Lys (2000b: 7). For an excellent overview of the value relevance literature, see Holthausen and Watts (2001); for an overview of value relevance studies that are especially relevant for CCV purposes, refer to section 5.2.1.

the stock price (i.e. assuming that the bases of reference do not fully explain the stock price) yields the following slight change to the basis equation:

$$PRICE_t = \sum_{n=1}^{x} a_{nt} \cdot BR_{n\tau} + \eta_t \qquad (3.6)$$

Since accounting figures are usually used as bases of reference when performing CCV, the similarity of this equation to the *value relevance equation* becomes obvious. As a consequence, for purposes of determining the relationship between bases of reference and stock prices within the framework of CCV, research clearly has to focus on the aspects of value relevance. From the point of view of CCV, information content and valuation relevance aspects are not very interesting since they concentrate on explaining expected returns. The aim of CCV, however, is to determine corporate values or potential market prices, not to predict future returns.[97] Additionally, information content studies focus rather on the set of accounting figures as a whole than on single variables.

Equation 2.4 (which is stated without an error term) is based on the assumption that the basis of reference (or bases of reference respectively) can fully explain the corporate value. However, it remains to be proven which bases of reference really explain how much of the stock price. To put it differently, classical CCV assumes a perfect value relevance of the variables included in the model. To accurately perform CCV it is thus necessary to include only those accounting variables that really have a high degree of value relevance. It is important to note that there is a close relationship between the quality and the value relevance of accounting figures. A theory about value relevance requires the existence of a high quality of accounting variables; otherwise all theoretical predictions would fail in practice.

The aim of this part of the study is not to measure value relevance (this is done in section 5.2). It is rather to develop economicly sound theories about which degree of value relevance certain bases of reference should have under certain conditions. This issue could best be addressed by considering that price, resp. value, is calculated by discounting future financial benefits (Equation 2.1.). Generalising Equation 3.6 and substituting price by the expression of Equation 2.1 leads to the following basis equation for theoretical analyses:

$$\lim_{m \to \infty} \sum_{n=1}^{m} E\left[FB_{t+n}\right] \cdot (1 + c_{t+n})^{-n} = f(BR_{1\tau}; \ldots; BR_{x\tau}; \eta) \qquad (3.7)$$

Similar to Equation 3.5, the lower the influence of η, the higher the value relevance of the set of reference variables BR is. Moreover, it becomes obvious that the set of reference variables has high value relevance if these variables have a high predicting power for future financial benefits and future discount rates. To put it differently:

---

97 As a matter of course, these two goals are not totally independent, since corporate values are the result of discounting future (expected) returns. Nevertheless, CCV focuses clearly on the determination of values, not on the prediction of values.

Value relevance is a function of how well the chosen reference variable(s) is/are an indicator for future financial benefits and discount rates.

The theoretical examination of value relevance uses this notion to determine under which circumstances a certain (set of) reference variable(s) is expected to be highly value relevant and under which circumstance it is not.[98]

When choosing a certain basis of reference, appraisers face four questions that are of interest: (1) Should we value the firm as a whole and subtract the market value of debt afterwards, or should we directly value the equity part of the company? (2) Should we use performance-oriented bases of reference (such as earnings or cash flows) or substance-oriented bases (such as the book value of equity)? (3) Should we use accrual bases of reference or cash flow-oriented bases of reference? (4) Should we use trailing bases of reference or forward bases of reference? The following sections shed light on these problems and try to provide answers to the questions from a theoretical point of view.

3.2.1.2 Enterprise Valuation Versus Equity Valuation

Not only in CCV but rather in every business valuation approach, appraisers face the question whether they should perform enterprise valuation or equity valuation. Although both variants lead to the same result[99] – the value of equity – their proceeding is different. The enterprise approach calculates the value of a company's equity indirectly as the value of the whole firm less the value of debt. Contrary to that, the equity approach directly values the part of the company that belongs to the equity investors. While in some – mainly Anglo-Saxon driven – text books about valuation the enterprise valuation approach is regarded to be generally advantageous[100], others do not favour one method over the other (see e.g. Schultze,

98 This approach is similar to that of some modern dynamic stock valuation models that have emerged from the Ohlson framework (see Ohlson 1995) during the last years; see e.g. Berk et al. (1999), Ang and Liu (2001). These models try to explicitly take into account the time series behaviour of variables that influence the stock value. However, all models have in common that they are based on some sort of stochastic processes whereas the approach used in this paper clearly focuses on the economic influence. Moreover, due to mathematical restrictions the above-cited models often fail to accurately value stocks with a positive probability of realizing zero or negative earnings in the future, see Dong and Hirshleifer (2004). The approach used in this study is not associated with such problems.

99 This is the case as long as changes in the capital structure are consistently considered in both approaches, see Schultze (2003: 318-348); IDW (2002: 110); Damodaran (2002: 399); Damodaran (2001b: 775); Copeland et al. (2000: 131-132).

100 See e.g. Copeland et al. (2000: 132-133). Their reasoning is that this approach helps in identifying separate sources of value of a company, aids the search for value creat-

2003: 484). Anyway, the enterprise method is more widespread among appraisers (see Kames, 2000: 68, 71; Wichels, 2002: 146, 148).

Fig. 9. Reference variables for enterprise and equity valuation

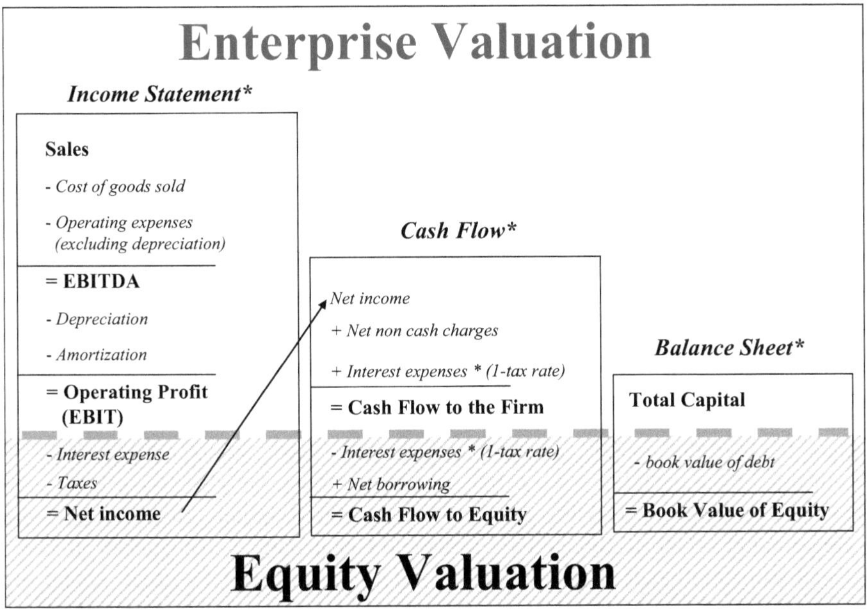

*Assumptions: Simplified income statement and balance sheet under US-GAAP, no discounted segment, no extraordinary items, no effect of changes in accounting principles; the presented cash flows are just two multiple possible cash flow variants.

Source: own figure according to Seppelfricke, 1999: 305; Herrmann, 2002: 76.

From a technical point of view, the difference between enterprise valuation and equity valuation in CCV lies in the nature of the bases of reference. For enterprise valuation, the bases of reference used should belong to both equity and debt capital providers, while for equity valuation the bases of reference only belong to the equity investors.[101] To put it more precisely for accounting figures: Starting with sales and going down the income statement, the accounting figures have to serve as bases of reference for enterprise valuation as long as the claims of debt providers are not satisfied. This is the case, as a rule, for all bases of reference that are calculated before considering the interests paid to creditors. As regards substance-oriented accounting figures, the book value of total assets (total capital) serves as an enterprise valuation reference variable. Consequently, the book value of equity

ing ideas and can be applied consistently at different valuation levels (whole firm level or individual business units level).

101 "If the numerator for a multiple is an equity value, then the denominator should be an equity value as well. If the numerator is a firm value, then the denominator should be a firm value as well" (Damodaran, 2002: 457).

represents an equity valuation reference variable (since the claims of debt investors are subtracted from the book value of total assets).

Similar is true for cash-oriented bases of reference. As long as the cash flow goes to all capital providers, it has to be used as a basis of reference for enterprise valuation. In contrast, if the cash flow just goes to equity investors, it has to be applied as a basis of reference for equity valuation. Figure 9 illustrates the appropriateness of different variables for either equity valuation or enterprise valuation.

At this point, some short remarks on a widespread misunderstanding of this principle are necessary. A single-factor model often used in valuation practice is the price/sales ratio (see Damodaran, 2002: 544-548; Kames, 2000: 100-103; Fernandez, 2002: 3; Stowe et al., 2002: 216-222). However, as can be seen from Figure 9, the use of this ratio cannot be justified economicly. The variable "sales" clearly belongs to all capital providers and therefore has to be set into relationship to the enterprise value of a company. If that is not done, the influence that different corporate financing structures have on the equity value cannot be properly accounted for. More precisely, the equity of a company that carries a big amount of debt relative to the average comparable company will be overvalued because a bigger portion of sales is financed with debt. The opposite is true for a company that has lower levels of debt than the average peer group company. The following example highlights the misvaluation using the price/sales ratio (see Figure 10).

In this illustration the target company and the comparable company both have the same amount of sales. Therefore it is reasonable to assume that they have the same amount of capital available to finance their business. Consequently, it is assumed that the market value of total capital equals 100 in both cases. Using the price/sales model, the market value of equity of the target company is calculated as 100, independent of the amount of debt financing. However, using the EV/sales ratio, the market value of equity VE decreases with increasing debt financing (i.e. $VE = SALES \cdot \left(EV / SALES \right) - Market\,Value\,Debt$). Following the price/sales ratio, the amount of debt has no influence on the market value of equity. This, in turn, would mean that a company has the riskless opportunity to increase the equity value by simply boosting sales via debt financing of operational activities. It would also mean that a company could substitute equity financing by debt financing without changing the market value of equity. Obviously, these two scenarios are not consistent with economic reality.[102]

Irrespective of this short discussion about the pointlessness of the price/sales ratio, the central question remains: Under which circumstances are enterprise value related bases of reference more value relevant than equity value related bases of reference?

---

102 A similar opinion is provided by Schwetzler and Warfsmann (2005: 50-51). Note: Of course, companies with totally different capital structures should not be seen as comparable companies when using sales as the reference variable because of the impact of leverage on the cost of capital, see Copeland et al. (2005: 564-570, 605). However, the price-sales ratio does not deliver proper valuation results for minor differences in the capital structure, either. The numbers in the example are chosen to better illustrate the pointlessness of that ratio.

Fig. 10. The pointlessness of the price-sales ratio

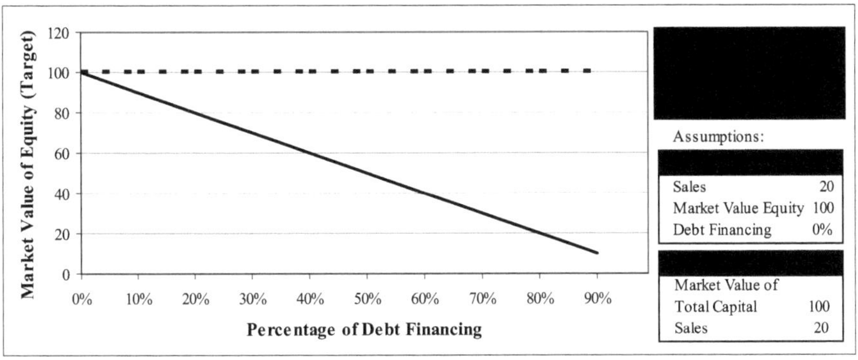

Source: ZEW

To answer this question, a step-by-step analysis is necessary. In the first step, the two bases of reference, operating profit (EBIT) and net income, are compared to each other. Assuming a company with a stable capital structure and a stable operating business, there should be no major differences between the relevance of EBIT for the enterprise value and the relevance of net income for the equity value. Obviously, both accounting variables are rather good predictors for the respective future financial benefits that are typically used in direct valuation settings (i.e. free cash flows). Assuming that these cash flows remain stable over time, the close connection between the reference variables and the financial benefits is shown in Figure 11.

In the case of a stable capital structure, there are no fluctuations in net income due to financial activities, and no net borrowing, either. This means that both bases are closely connected to the respective free cash flows since non-cash charges partly offset capital investments in a stable operating business. Net income possibly has slightly higher value relevance because the impact of taxes has already been considered in this accounting figure.

If, however, a company faces high capital structure volatility, then EBIT is expected to be more value relevant than net income. This is the case because the gap between net income and free cash flow to equity (FCFE) widens due to the impact of net borrowing as the financial activities vary over time. In contrast, EBIT and free cash flow to the firm (FCFF) are not at all affected by an unsteady capital structure. This is largely consistent with the general conclusion that in direct business valuation, the equity approach is said to be easier and more straightforward to use when the capital structure of the company is not expected to be volatile in the future and if the company does not have a significant amount of debt outstanding (see Stowe et al., 2002: 116; Damodaran, 2002: 399). These outlines also principally apply to the relative value relevance of book value of equity and book value of total capital, as well as to the value relevance of cash flows to equity and cash flows to the firm.

Fig. 11. The translation of reference variables to respective free cash flows

Basis of Reference	*Respective Financial Benefits in Direct Valuation*
Sales	→ **Free Cash Flow to the Firm (FCFF)** → (Sales - COGS - operating expenses)*(1-tax rate) + Net Non-Cash Charges - Working and Fixed Capital Investments
EBITDA	→ **Free Cash Flow to the Firm (FCFF)** → EBITDA*(1-tax rate) + Net Non-Cash Charges*(tax rate) - Working and Fixed Capital Investments
EBIT	→ **Free Cash Flow to the Firm (FCFF)** → EBIT*(1-tax rate) + Net Non-Cash Charges - Working and Fixed Capital Investments
Net Income	→ **Free Cash Flow to Equity (FCFE)** → Net income + Net Non-Cash Charges + Net borrowing - Working and Fixed Capital Investments

Source: ZEW

The comparison of reference variables that are both on the enterprise level is not quite so straightforward. The relative value relevance of EBITDA and EBIT clearly depends on whether depreciation and amortization reflect the true economic value of periodical cost allocation. If they do, then EBIT is more value relevant because – as can be seen in Figure 11 – it is very similar to FCFF in a stable business environment. If, however, they do not, then EBITDA is expected to be more value relevant since it is less distorted by accounting discretion and therefore is economicly more plausible. An increasing timely instability of the operating business that is associated with a timely instability of future cash flows decreases the value relevance of all bases of reference. However, no final statements can be made on how fluctuations in the operating business affect the relative value relevance of the two variables discussed here.

As compared to EBIT and EBITDA, sales should have lower value relevance because their similarity with the respective cash flows depends on additional items, such as costs of goods sold (COGS) and other operating expenses (see Figure 11 above). To put it more precisely, sales can only be expected to have the same value relevance as EBIT and EBITDA if costs represent the same percentage of sales every year. This is especially not the case for companies with a high portion of fixed costs relative to variable costs and varying operating business, because then total costs are expected to fluctuate heavily with the cash flows.

3.2.1.3 Performance-Oriented Bases Versus Substance-Oriented Bases

A second very important question referring to reference variable selection is whether to use performance-oriented bases of reference (such as cash flows, net income, EBIT, EBITDA and other accounting figures from the income statement) or substance-oriented bases of reference (mainly the book value of equity and total assets). In CCV practice both kinds of bases enjoy certain popularity (see Kames, 2000: 100-103; Wichels, 2002: 146, 148; Fernandez, 2002: 3). However, in order to decide under which circumstances one or the other group of accounting figures is more appropriate, a closer look at their relative value relevance is necessary.

The case of reorganisation

Often discussed in substance- versus performance-orientation is whether the target company (and consequently the comparable companies, too) is expected to continue its current business activities. In direct valuation approaches the focus on substance is more likely the more a company faces an operating change. That means on the one hand that companies that are expected to abandon – or that already have abandoned – their operating business (e.g. due to bankruptcy) are valued by applying asset-based approaches. On the other hand, asset-based methods are also – at least partly – appropriate if companies change current business activities by adapting its resources to different uses.[103] In the case of abandonment and sale of assets, the company is worth exactly what it is expected to get for the sales of net firm resources. In the more complex settings of strategy change and employment of assets in alternative uses within the firm, the value of the company depends on the specific use to which firm resources should be adapted.

The factual break in current business activities is, however, not a necessary requirement for the substance focus. Asset-based approaches also have some importance (e.g. as a supporting tool) when there is a certain probability for the change of operating business. This is especially the case for low-income companies, i.e. companies that do not earn the normal rate of return on capital with their current business activities (see Leuner, 2002: 649-670). In the context of CCV this means that the value relevance of substance-oriented bases of reference grows with an increasing probability of reorganisation. To put it differently, asset-based reference variables are more value relevant for fundamentally underperforming companies than for fundamentally outperforming companies (see Jan and Ou, 1995; Burgstahler and Dichev, 1997; Barth et al., 1998; Subramanyam and Venkatachalam, 1998; Yee, 2000; Chen and Zhang, 2002).

[103] See section 2.2.3.4 for a short description of when to use the different kinds of asset-based valuation approaches.

The case of going concern

However, the value relevance of asset-based reference variables is not only driven by (potential) abandonment or adaptation of business activities. These variables are also likely to be value relevant in case a company continues its current business. In order to compare the value relevance of the two categories of accounting figures – asset-oriented or performance-oriented ones – in the going concern case, an analysis of different valuation circumstances is necessary. In fact, for appraisers there are certain indicators that determine which kind of ratio is supposed to deliver more accurate results in a practical valuation setting, i.e. which of the two categories is more value relevant. These indicators can be divided into three dimensions.

The first dimension is the availability of capital. Usually there are two possibilities of financing new investments: either by internal capital sources (retained earnings) or by external capital sources (debt). If a company has no access to the debt market, corporate growth is possible only if current and future investments are profitable, i.e. if current and future earnings are positive.[104] Consequently, the amount of current earnings well defines future growth possibilities. In this case earnings have high relative value relevance as compared to book value.[105]

In contrast, current earnings are not necessarily a good indicator for future performance if a company can also rely on external finance. In this case the growth rate still depends crucially on the return on equity for future investments but no longer exclusively on the retention rate and therefore no longer exclusively on current earnings. However, in this case it is not clear whether book value is more value relevant than earnings because access to external finance usually still depends on the performance of a company and therefore on its amount of earnings. Thus, it can only be stated that from a financing perspective there are some reasons to consider book value as not irrelevant for the valuation of companies and the relevance is expected to be higher (relative to the relevance of earnings) for those companies that have good access to external capital.

The second dimension is the value added process of the company (i.e. the depth of production, the production and capital intensity). At one extreme of this dimen-

---

104 This becomes clear when looking at the determinants of the sustainable growth rate of earnings g: $g = ROE * earnings\ retention\ rate$, see Solnik and McLeavy (2004: 286). A lack of external financing causes a decline of the growth rate since debt financing is necessary to keep the capital structure constant, see Brealey and Myers (2003: 837) (in this context it is important to note that the comment in Benninga and Sarig (1997: 318): "if the assumption that growth is exclusively financed by retaining some earnings is wrong, this way of estimating growth [see the equation above in this footnote, the author] will not work." is definitely false). Moreover, restricted access to debt financing gives rise to a decline in financing flexibility, which is especially a problem if the company faces an unsteady project structure. Both arguments emphasize the importance of current earnings for future value creation of the company.

105 The focus here is on earnings in general, not on retained earnings. This is done because the amount of earnings crucially determines the amount of retainable earnings. See also section 3.1.1, Footnote 58, regarding the irrelevance of the dividend policy in accounting regimes with and without the clean surplus relation.

sion we find a group of companies that does not employ a significant amount of balance sheet assets during the value added process or that does not have a high proportion of variable assets, such as the services industry. Future cash flows of these companies do not depend crucially on the book value of equity. In this case current performance (e.g. current earnings) is a good indicator of future performance. Thus, it is earnings rather than book value that are value relevant.

On the other extreme we find a group of companies that employ a high proportion of balance sheet assets during the value added process (asset heavy companies) or that have a great amount of variable assets relative to total assets. Examples of these companies are utilities and banks. In this case, book value typically is a good indicator for the amount of firm resources, i.e. the firm can be viewed as a collection of separable assets whose collective amounts are good estimates of the market value of the company. For these companies, future income earned is on average in close relation to the book value and, thus, book value is of relatively high value relevance.[106]

The third dimension is the state of the industry. The degree of competition in an industry determines the level of profitability of companies and the possibilities to earn abnormal earnings in the long-term. In pure competition markets (price taker markets) economic profits approach zero in the long run – i.e. companies earn a normal rate of return.[107] However, in most industries the assumption of pure competition is not satisfied. Porter provided a framework which points out the different forces that influence the real degree of competition of an industry (see Porter, 1980: 3-29; Porter, 1998: 21-34). These five forces are: the rivalry among companies, the buyer power, the supplier power, the barriers to entry and the threat of substitutes.

In industries that are only weakly influenced by these industry forces (with the exception of the barriers to entry which are assumed to be high[108]) companies can earn long-term abnormal profits. A low rivalry permits strategic moves such as changing prices or differentiating products. In industries with low degrees of buyer and supplier power[109] companies have the power themselves to improve

106 See Peemöller et al. (2002: 207); McCarter and Aschwald (1992: 153); this is especially not the case for companies with a high proportion of intangibles such as in the pharmaceutical industry or the IT industry.

107 See Gwartney and Stroup (1997: 515). Even in note purely competitive markets *ROE* reverts to an average industry or economic wide level, see Lev (1969), Penman (1991), White et al. (1997: 191). However, in these cases reversion takes a longer time. The focus on industry structures as the main determinant of profit persistence goes back to the neoclassical perspective, see Mueller (1990: 2); company characteristics might also have an impact on the possibility to earn economic rents (this is the so-called *Schumpeterian View*) but it is assumed here, that this is only of importance in the short run.

108 If below the influence of industry forces is called "weak", barriers to entry are always assumed to be high and vice versa.

109 A high intensity of production (high depth of production) typically leads to a low dependence of suppliers and therefore to a low supplier power. Thus, the second and the third dimension are not totally independent.

their profit-cost relationship and therefore their margins. When barriers to entry are high, the degree of rivalry will stay low. Thus, companies can keep or even increase their abnormal profits. Finally, a low number of substitute products or the non-existence of close substitute products raises companies' ability to increase prices.

But how does this affect the relative value relevance of earnings and book value? To answer this question, a closer look at two direct valuation models is necessary. The basic model that proceeds on the value relevance of book value and earnings is the RIV model. Starting from the assumption of the clean surplus relation, it can be shown that the market value of equity equals the sum of the current book value and the present value of future residual earnings (see Preinreich, 1938: 240; Lo and Lys, 2000a):

$$PRICE_t = EQUT_t + \lim_{N \to \infty} \sum_{n=1}^{N} \left(NI_{t+n} - c_e \cdot EQUT_{t+n-1} \right) \cdot \left(1 + c_e\right)^{-(t+n)} \tag{3.8}$$

where NI_{t+n} is the net income at time $t+n$ and ($NI_{t+n} - c_e \cdot EQUT_{t+n-1}$) are the residual earnings (residual income RI) at time $t+n$. In this model, book value serves as a proxy for expected future normal earnings (see Penman, 1992). However, the model is stated very generally. Thus, no explicit market value forecasts can be made solely on the basis of current accounting data.

Ohlson (1995) extended the RIV model by introducing a system of linear information dynamics, which postulates certain time series behaviour of residual earnings. More precisely, the market value of equity is expressed as a function of residual earnings, information not yet captured by accounting ν – both following an AR(1) process, i.e. a certain autoregressive process – and current book value (see Ohlson, 1995: 670-671):

$$PRICE_t = \left(1 - k\right) \cdot EQUT_t + k \cdot \left(\pi \cdot NI_t - DIV\right)_t + \lambda \cdot \nu_t \tag{3.9}$$

with $k = \omega \cdot r / \left(1 + r - \omega\right)$, $\pi = \left(1 + r\right) / r$ and

$$\lambda = \left(1 + r\right) / \left[\left(1 + r - \omega\right) \cdot \left(1 + r - \gamma\right)\right]$$

where r = risk free rate of interest, ϖ and γ are persistence parameters with $0 \leq \omega, \gamma \leq 1$.

Empirical studies that rely on the Ohlson model mostly show its superiority to other (classical) accounting based valuation models.[110]

The Ohlson model reveals that prices depend heavily on current book value if the persistence of current earnings is low (strong industry forces), and that prices rather depend on current earnings if the persistence of current earnings is high (weak industry forces). Thus, for companies that do business in a world of weak

110 See e.g. Richardson and Tinaikar (2004), Courteau et al. (2001). Dechow et al. (1999), and Myers (1999), found only weak support for this hypothesis.

industry forces the current performance (current earnings) is a good indicator for future performance. Outperformers can keep their strategic advantages because of their power while underperformers do not have the possibility to defend against the outperformers. In such a case, earnings should have relatively high value relevance. If, however, the five forces heavily influence the industry, long-term economic profits are driven to zero[111] or at least profitability becomes equal for all companies. Given that profitability can be expressed as a certain percentage of book value, then book value is relatively more value relevant.

Figure 12 visualises the determinants of the value relevance of book value and earnings. It combines all three dimensions and shows recapitulatorily when substance-oriented bases, such as the book value, and when performance-related bases, such as earnings, exhibit relatively high value relevance. It should be noticed that this economic approach to selecting valuation ratios basically applies to both the enterprise level (i.e. valuation of the whole company) and the equity level.

It is worth noting that in economic reality none of the scenarios which definitely recommend the application of one of the two ratios is satisfied. Thus, the position of a company can usually be interpreted as an intermediate point between the two extremes of each dimension. First, companies generally have the possibility to raise a certain amount of debt capital to finance new projects even though the amount of funds that can be raised varies over time because of changes in corporate liability management or dependent on conditions on capital markets. Second, regarding the production process of a company, usually neither the total amount of balance sheet assets nor none of the assets are employed in the value added process. Companies rather exhibit a certain degree of asset employment. Additionally, balance sheet assets are more or less a good proxy for the amount of firm resources[112], but do not correspond perfectly. Third, there is a certain degree of competition in every industry. However, pure competition industries are scarce. Even if some of Porter's industry forces strongly influence corporate behaviour, there is hardly a situation in which the entirety of forces controls the industry.

To put it differently, using the "Earnings – Book Value" relevance cube may demonstrate tendencies as to which accounting figure has the higher value relevance. However, in reality both variables presumably have a certain explanatory power – only the weightings are different for each company.[113]

---

111 If economic profits are zero then the company earns its cost of capital. This cost of capital can be assumed to be approximately equal for all the companies within the industry as long as there are no major differences in capital structures, see section 4.3.2.4.

112 This heavily depends on the applied accounting standard, too.

113 This complementary nature of book value and earnings is also supported by the FASB: "[The income statement; the author] can be interpreted most meaningfully ... only if it is used in conjunction with a statement of financial position, for example, by computing rates of return on assets or equity", FASB (1978b) paragraph 24a.

Fig. 12. The "earnings – book value" relevance cube for going concern companies

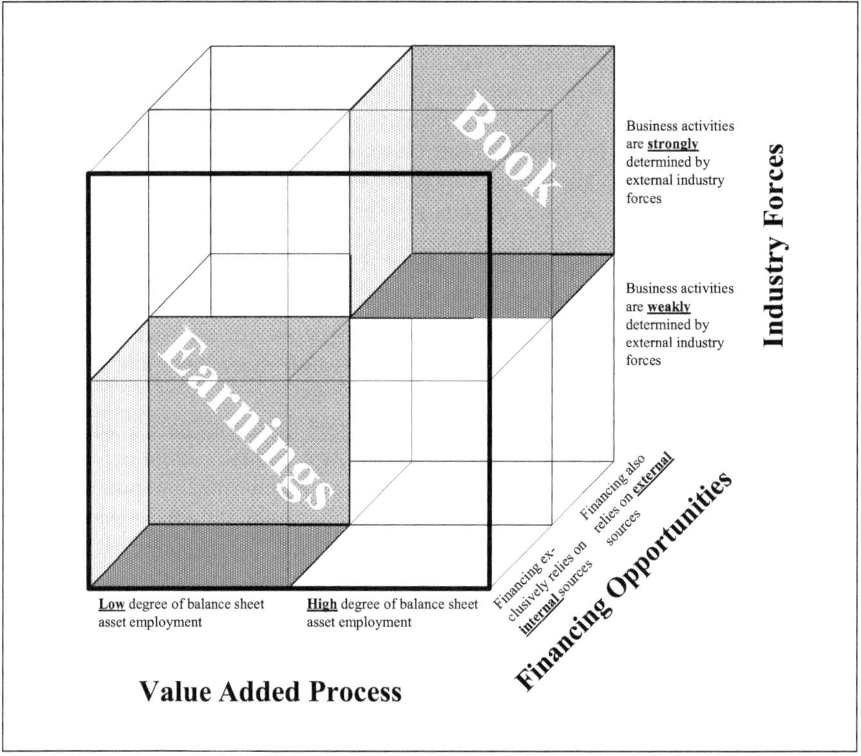

Source: ZEW

3.2.1.4 Accrual Bases Versus Cash-Flow-Oriented Bases

Appraisers typically face the question whether to use bases of reference, which are drawn from the financial statements, or whether to use cash-oriented bases. This question is only relevant when using performance-oriented bases because there is no cash-oriented equivalent for asset-based accounting figures such as, e.g., book value of equity or total assets. As from the perspective of direct valuation approaches, there is a clear theoretical international preference for cash orientation.[114] Furthermore, an international overview of direct valuation and capital budgeting models applied reveals that cash-oriented models are used more often by far than accounting based models (see Ryan and Ryan, 2002; Graham and Harvey, 2001). In Germany, however, things are different, i.e. earnings capitalisation

114 See exemplary Copeland et al. (2000: 73-87: chapter 5 "Cash is King"). The aversion towards earnings capitalisation models is also due to the possibility of accounting figures manipulation.

models enjoy widespread popularity. Particularly auditors and tax consultants prefer to use the “Ertragswertverfahren”.[115] One argument for their popularity is that the financial benefits (i.e. earnings) can be determined relatively easily based on prospective balance sheets and income statements (see Peemöller and Kunowski, 2005: 212).

In the context of CCV, however, the reasoning regarding the preference for one of the two financial benefits is not quite so straightforward. Certainly, accrual bases of reference are much easier to obtain. Trailing bases can be copied directly from the last reported financial statements and forward bases are usually available from certain databases (e.g. I/B/E/S consensus earnings estimates) for many listed companies. Easy obtainability, however, is not of major importance in CCV. To determine appropriate corporate values, appraisers should make every effort to get the data necessary for the application of the chosen valuation approach. Moreover, the calculation of cash flows is usually not an insuperable obstacle for appraisers – neither the calculation of trailing cash flows nor that of near-term future cash flows. As a consequence, easy obtainability would only be an issue if both kinds of reference variables lead to the same results anyway. As will be outlined below, this is not necessarily the case.

The assessment of the relative value relevance of earnings and operating cash flows can be done from two perspectives. *One of these perspectives* is the accounting quality. In the first instance, this approach requires a closer look at the composition of both financial variables. A major difference here is that earnings include non-cash accruals while operating cash flows do not (see White et al., 1997: 36-43). These accruals typically reflect changes in asset or liability values that do not result from a cash inflow or outflow. The changes, in turn, have to be regarded as an anticipation of future cash flows. Therefore, accruals are a means of allocating future expected cash flows to the current period (see White et al., 1997: 40). As a result, accounting earnings are assumed to better predict future cash flows than current cash flows do.[116] Consequently, earnings should have higher value relevance than cash flows under the assumption that management has good forecasting abilities and always behaves rationally. However, the prudent management assumption does not hold automatically. Some of these accruals are subject to management discretion and, consequently, there is room for income manipulation and earnings management. To put it differently: A clear statement regarding the favourability of either earnings or operating cash flows is as of now not possible. Instead, an assessment of the relative value relevance of these two financial figures requires an in-depth analysis of the quality of accounting.

---

115 See Peemöller et al. (1994: 741-749); Peemöller et al. (1999: 622); Peemöller and Kunowski (2005: 204-206). However, as outlined in section 2.2.3.1, the methodological differences between the equity approach of DCF valuation and the German form of the “IDW-Ertragswertverfahren” are rather small.

116 This opinion is shared by the FASB: “Information about enterprise earnings based on accrual accounting generally provides a better indication of an enterprise’s present and continuing ability to generate cash flows than information limited to the financial effects of cash receipts and payments.”, FASB (1978a), paragraph 44.

The second perspective is the life cycle of a company. Following this rather functional approach, there should be a difference between the relative value relevance of earnings and operating cash flow for new firms and for established firms. The theory goes back to the break-down of corporate value into assets and growth opportunities:[117]

Value of the firm = Value of Assets in Place + Value of Growth Opportunities

During early life cycle stages the focus of companies is clearly on growth opportunities. Thus, in these stages, the variable that provides more value relevant information is one that better explains the value of growth opportunities. Usually – compared to earnings – operating cash flows are a better mean for young companies to internally fund growth, because they are a true measure of the liquidity available (see Black, 1998: 9). Consequently, cash flows should be more value relevant.

Contrarily, the value of mature companies is to a large part determined by the assets in place. These assets generate stable revenues and expenses, and earnings can be expected to be more permanent. In such an environment, earnings are assumed to be more value relevant than operating cash flows (see Cheng et al., 1996).

Apart from these two perspectives, one important issue regarding the value relevance of cash flows has to be considered: Using free cash flows (both on an enterprise and on an equity basis) as a basis of reference is much more problematic than using operating cash flows. The reason for this is that it is not possible to differentiate between a low free cash flow due to a low operating cash flow on one hand (which is a bad sign) and due to high investments on the other hand (which is usually a good sign because of the future expected benefits of these investments) and vice versa (see Penman, 2003: 203). Thus, appraisers cannot draw any clear conclusions from single free cash flows – these financial figures are not useful in predicting future cash flows. Consequently, free cash flows have to be seen as a very inaccurate basis of reference from a theoretical point of view.

To summarise: A general transferability of the Anglo-Saxon direct valuation ideology (i.e. the cash focus) to CCV is not appropriate.[118] It is rather important to analyse the accounting quality of accruals and to consider the state of maturity of the company in order to judge the relative ex-ante value relevance of earnings and cash flows (see Figure 13). Free cash flows, however, are not at all appropriate

---

117 See Myers (1977). This break-down must not be confused with the break-down of the corporate value into book value and earnings as it is done in the Ohlson model, although both approaches are familiar.

118 Therefore the following comment in Richter and Herrmann (2002: 5) is wrong: "In apparent contradiction to basic principles of valuation they [i.e. Liu et al. (2002); the author] find that multiples based on forecasted earnings clearly outperform different kinds of cash flow multiples."

bases of reference because of the cash flow decreasing impact of certain capital investments.

Fig. 13. The "earnings/cash flows" relevance square

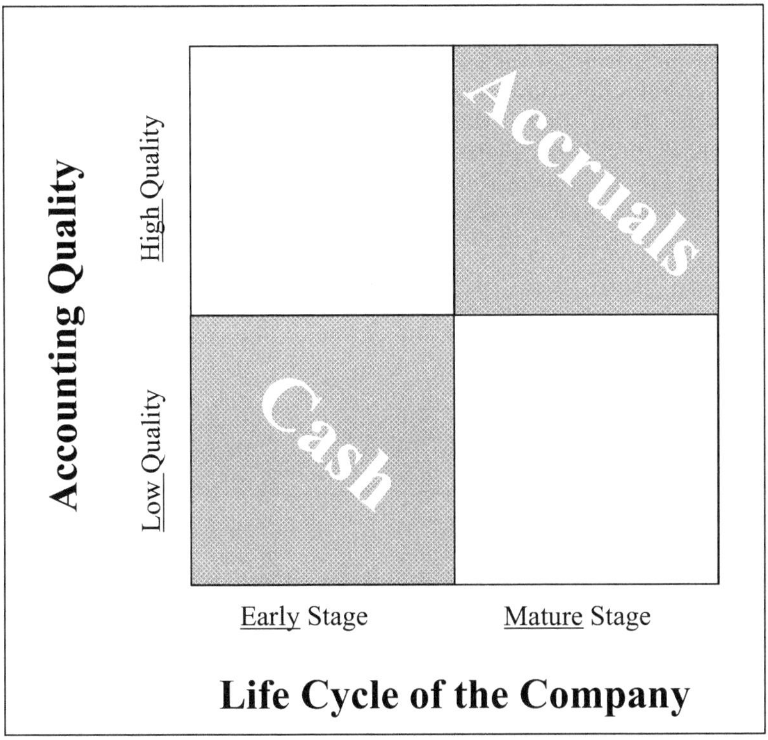

Source: ZEW

3.2.1.5 Forward Bases Versus Trailing Bases

The choice between forecasted future (forward) and contemporarily reported (trailing) bases of reference is also of particular interest in CCV. Typically, this issue is relevant only for performance-oriented bases of reference (like earnings or cash flows). Substance-oriented bases – such as book value of equity or total assets – are seldom forecasted. One reason for this lack of forecasts is that the percentage change between trailing and forward book values is rather small and therefore a possible utility gain of the forecast is typically small. Another reason is that the estimation of these variables not only requires a forecast of the operating and financing activities of a company, but also of the dividend policy.

In general, appraisers have to balance the advantages of the higher (forward looking) amount of information that is inherent in forward bases against the disadvantages of the uncertainty associated with these forecasts. Typically, the uncertainty about the forecast validity decreases with the future reporting date approaching because of the shorter time horizon: Appraisers have more information about future accounting variables and the risk of unexpected events has decreased. Thus, the value relevance of forward bases increases with a decreasing time to the reporting date. Additionally, the information inherent in trailing bases is getting old and therefore less value relevant the longer the time period from the historical reporting date. This effectively means that the use of forward earnings is more advisable from the perspective of value relevance the closer the future reporting date is. Figure 14 illustrates this. However, if the appraiser has access to a sufficiently large and high-quality set of information, the use of forward bases could be superior to trailing bases from a value relevance perspective even at or shortly after the date of reporting. For the same reason, the use of forward bases other than for the next period (e.g. 2-year-forward bases) might also be appropriate under certain circumstances.

Fig. 14. The general value relevance of trailing and forward bases of reference

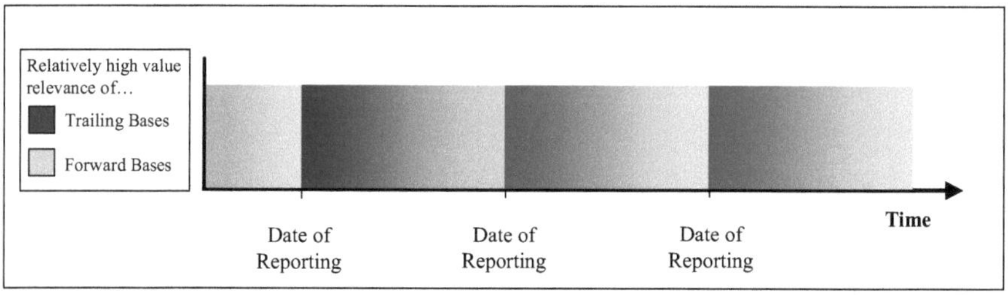

Source: ZEW

Especially with regard to earnings[119] it can be theoretically shown that forward earnings-value relations are less demanding on the accounting system than trailing earnings-value relations are. More precisely, the probability of linear earnings-value relations over time – and therefore of high value relevance – is higher for forward earnings than for trailing earnings. This result holds especially in the realistic case that market participants cannot observe all information variables (see Ohlson, 1991; Yee, 2004). The reason for this phenomenon is that earnings forecasts naturally neglect unexpected non-recurring items (so-called “one-timer”). Therefore they are typically more sustainable than trailing earnings.

However, the problem of uncertainty about future events remains and, thus, no general statements about the relative value relevance of both kinds of earnings can be made. These theoretical studies only indicate that the accounting quality is usually higher for forecasted than for reported earnings, all other things being equal. Moreover, as outlined in chapter 2.4.1, appraisers are typically aware of the problem with non-recurring items. Consequently, appraisers might adjust current earnings for CCV purposes so that they better resemble the economicly reasonable figures.

119 Earnings are by far the most often forecasted accounting figure. Therefore the bulk of all empirical studies dealing with forward accounting variables focuses on earnings.

Part of the recent empirical research suggests that trailing (i.e. current) earnings do not reflect the underlying economic events in a timely manner (see Basu, 1997; Easton et al., 2000). Following these studies, a portion of earnings is already anticipated by the capital market before the announcement date. Event studies focusing on the relative information content of current earnings and management forecasts typically find that stock prices react stronger to earnings estimates announcements than to reported earnings announcements (see e.g. Conroy et al., 2000). They conclude that forecasts are far more pricing relevant than current variables. A more differentiating study found that the market places relatively more importance on earnings forecasts if the company is in a high growth stage or if the company discloses earnings surprises (see Ota, 2001). This phenomenon is due to the lack of current earnings sustainability in such cases, which, in turn, reduces their predictive ability for future expected financial benefits.

Another important issue – when dealing with forward earnings – is that appraisers who do not conduct their own forecasts but plan to use forecasts published by an external source have to act with care. Since the typical way forecasters reach their results is not disclosed, the automatic reliance on such a black-box-figure can easily yield noisy valuation results. Thus, it might be appropriate to challenge the formation process of these forecasts before using them. Such a review, however, is a difficult task. It can only be accomplished by recalculating earnings estimates step-by-step – which is in turn identical to an own forecast. While this proceeding might add to "remove some of the mystery surrounding analysts' forecasts"[120], it is not advisable for practical use. Typically, the quality (and the value relevance, too) of aggregated consensus forecasts (e.g. I/B/E/S earnings estimates) is quite high, even if they tend to slightly overestimate the real numbers (see Ababarnell, 1991; La Porta, 1996; Brown, 1996; Brown, 1997; Brown, 1998). Because of their broad acceptation at capital markets, these consensus forecasts can be seen as a proxy for the forecasts inherent in stock prices. Consequently, they can be expected to exhibit relatively high value relevance. In contrast to that, individual forecasts of single analysts or banks have to be expected to be of lower value relevance because of the high risk of unsystematic over- or underestimations.

3.2.2 Future Similarity Between the Target Company and Comparable Companies

The future similarity of the target company and the peer group crucially determines the choice of the basis of reference. This is not an issue of classical value relevance as described in section 3.2.1, where reference variables were chosen based on how they are presumably related to the market value of the target company. It is rather an issue of indirect cross-company value relevance. That means, reference variables should also be chosen based on the target company's future

120 Beaver (1999: 41). Regarding the amount and quality of information captured in analysts forecasts, see Cheng (2004).

development in relation to comparable companies, and based on the comparable companies' development in relation to the target company. This issue is mainly related to the choice between different bases of reference that stem from the income statement.

If the target company is likely to change its capital or cost structure such that it resembles that of the comparable companies in the future – assuming that comparable companies fulfil the selection requirements of section 3.1.1 – it is rather appropriate to choose the (trailing) bases of reference that are largely unaffected by these changes. More precisely, for a company that plans to change its portion of debt financing to the average peer group level, current net income will probably not be an appropriate basis of reference because it is affected by the interest paid on current debt. Consequently, the use of variables that are determined without considering financing costs (such as EBIT, EBITDA or sales) is more reasonable. Accordingly, a company that plans to adapt its cost allocation policy (i.e. the depreciation and amortization policy) to the peer group level should be valued using bases of reference that are determined before considering cost allocation (such as EBITDA and sales). Similarly, if a company will adjust its cost structure so that it corresponds to the average of the comparable companies, the use of variables that include current cost will most likely result in distorted valuation results. With regard to income statement variables only sales meets the demand of being unaffected by costs. Figure 15 illustrates the relation between future similarity of companies and choice of the basis of reference for two typical scenarios. Company *A* temporarily has lower depreciation than the peer group but these differences will even out in the future. Since EBIT and net income are distorted by these interim differences, appraisers should rely on EBITDA or sales. Company *B* temporarily has higher interest expenses and therefore the use of net income as a basis of reference will lead to biased valuation results.

This whole explanation is similar to that of the relative value relevance of enterprise value variables and equity value variables. The big difference is that this section also deals with the case where comparable companies change their cost structure or capital structure while the target company's financial relations stay constant. Thus, even if a certain basis of reference has high value relevance, it could be more appropriate to use other bases of reference in CCV because of the expected development of comparable companies. The scenario where comparable companies change to the target company level is, however, of minor importance in practice since usually such companies cannot be seen as "comparable" and therefore have to be removed from the peer group. One situation where this issue is relevant is the valuation of an established company within a young industry. In this case, peer group companies presumably change their capital structure or cost structure over time while the target company's balance sheet and income statement figures are supposed to remain relatively stable.

To summarise: Apparently, the provision for the future similarity of the target company and comparable companies is especially important when valuing companies that have certain different current characteristics than the set of comparable companies but are likely to become similar over time. In practice, this could be the case e.g. for distressed companies which face a debt restructuring or companies that were subject to a leveraged buy-out recently (changing capital structure over time in both cases) or for young companies (changing cost structure over time). It is, however, also an important issue for the valuation of rather established companies in a young industry.

Fig. 15. The relationship between future similarity of companies and reference variable selection in two typical cases

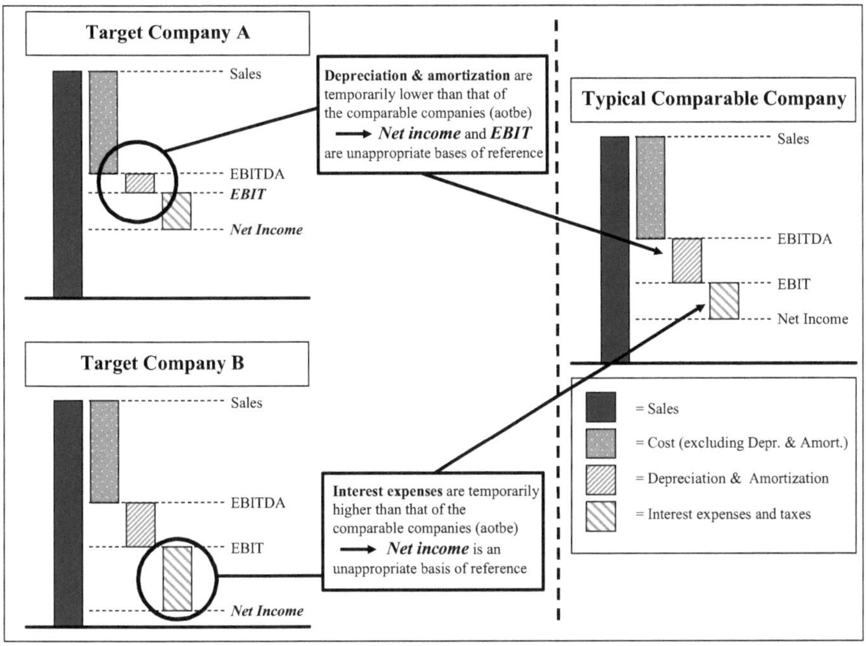

Source: ZEW

3.2.3 Technical Limitations of Valuation Models

Regardless of any economic reasonability (such as value relevance of the reference variables or the provision for future similarity of the target company and the comparable companies) some models are simply not applicable under certain circumstances because of technical limitations. These limitations are always related to the way in which data from the comparable companies are aggregated. The technical applicability, in turn, crucially determines the choice of the valuation model in general and especially the selection of the basis of reference. These limitations are described below for both general variants of aggregating the peer group results: the simple average approaches and the regression approaches. Additionally, the consequences for the reference variables selection are briefly discussed.

Limitations of models using the simple average approach to aggregate peer group results

The simple average approach to aggregate peer group results comprises the application of the arithmetic mean of the multiples, the median of the multiples and the harmonic mean of the multiples. These aggregation approaches are typically applied in the context of classical single-factor CCV models such as the PE ratio or the PB ratio (see Pratt et al., 2000: 243-244; Damodaran, 2002: 458-460). All these simple average approaches have common ground in that they result in a zero value of the target company if the reference variable has a zero value.[121] Moreover, these approaches yield negative corporate values if the basis of reference is negative.

The latter phenomenon is obviously not in accordance with the fact that equity values cannot become negative. This non-negativity of equity values holds because shareholders' liability is limited to the positive amount paid for the shares (see Gerke and Bank, 2003: 385). To put it more clearly: If the value of the whole firm would be equal to (or even less than) the value of debt (i.e. the value of equity would be zero respectively negative) the corporation would immediately default on the debt and, thus, would go bankrupt. This mechanism effectively prohibits the value of equity from being negative in practice.[122]

As a consequence, CCV models that rely on some sort of the simple average aggregation method are reasonably applicable only if the basis of reference is zero or positive. This non-negativity restriction typically has to be construed extensively; i.e. not only negative but also low positive reference variables should be excluded from these simple average aggregation models. Nonetheless, from a pure technical point of view, the only requirement is that the value of the reference variable must not be below zero.

This restriction is of minor practical relevance if substance-oriented bases of reference are applied since they rarely become negative. Contrary to that, performance-oriented reference variables such as cash flows or accounting figures drawn from the income statement can easily fall below zero. In practical valuation settings, appraisers react to that problem by simply ascending the income statement until they find an appropriate positive reference variable or, alternatively, try to find some sort of cash flow that is positive.

To put it more precisely for variables drawn from the income statement: Net income (which is located in the lowest position of the income statement) is calculated after subtracting all costs from sales. EBIT is calculated before the result from financing and before taxes. Since for non-financial companies in most cases interest expenses exceed interest income, EBIT typically exceeds net income. Therefore, compared with net income, it is more likely that EBIT is a positive figure. The probability of EBITDA being positive is even higher than the probability

---

121 See Figure 2 on page 41.

122 Consistent with this non-negativity restriction, the market value of equity equals the value of a long call option on the value of the whole firm with the face value of debt as the strike price, see Black and Scholes (1973); Damodaran (2002: 817-827); Copeland et al. (2005: 206-208).

of EBIT being positive, since EBITDA is determined before considering depreciation and amortization. Sales are always positive since they are not diminished by any costs.[123] Figure 16 highlights the use of different income statement related bases of reference in classical single-factor CCV models for four different scenarios.

Fig. 16. The use of income statement related bases of reference in classical single-factor models for four typical sales-cost scenarios

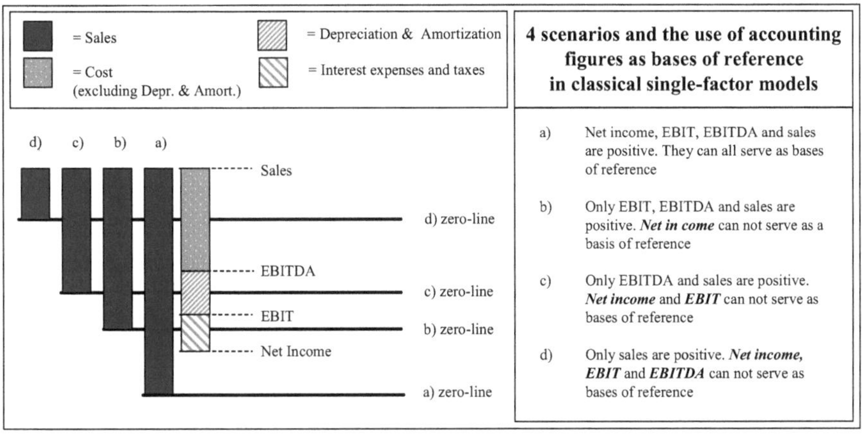

Source: ZEW

Limitations of models using the regression approach to aggregate peer group results

In most cases, regression based approaches in CCV are affected by typical small sample problems. Consequently, the central limitation to these models is the number of observations as compared to the number of estimated parameters. While models that are based on the simple average approach are principally applicable with only one comparable company, ordinary least square (OLS)-regression based models require at least as many observations as estimated parameters (see DeFusco et al., 2001: 294-295, 435). That means that a model which is based e.g. on earnings and book value (and therefore has three parameters including the intercept term) requires at least three comparable companies.[124] If additionally the quality (i.e. the goodness of fit) of the regression needs to be determined, one more ob-

[123] See section 5.2.4.1 for descriptive statistics about the frequency of positive net income, EBIT, EBITDA and sales.

[124] By suppressing the intercept term the number of comparable companies can be reduced by one. However, there is typically no economic reason for this suppression. In this context, it is worth noting that simple linear regression models with a zero intercept are very similar to models using the simple average aggregation method.

servation is necessary.[125] If regression approaches other than OLS should be applied still more observations are necessary.

There are two other problems of small samples that apply to regression based CCV models but also to simple average based CCV models. First, the smaller the sample the more vulnerable the regression to outliers is, i.e. to company data that are far away from the rest of the data. Second, the quality of the regression increases with the number of observations.

The regression based CCV models sometimes violate other assumptions made with the OLS regression approach.[126] In particular, the variance of the residuals might not be the same across all observations (heteroscedasticity) (see Kothari and Zimmermann, 1995: 168), and the independent (fundamental) variables might not be uncorrelated (multicollinearity) (see Herrmann, 2002: 120). However, these problems do not automatically lead to the inapplicability of CCV. In fact, to what degree these possible violations influence the regression result has to be analysed in the context of concrete valuations.

Sometimes it is mentioned that the relationship between corporate values and fundamentals should be non-linear rather than linear (see Herrmann, 2002: 120). This, however, is not a typical problem of the regression approach; models using the simple average aggregation approach are based on an even more severe underlying assumption regarding the linear relationship. Moreover, CCV models *can* be modified to account for a possible non-linear relationship.[127]

The argument sometimes heard, that regression based models violate the value additivity principle[128], is not true. In fact, the parameters are totally different from single projects or value components. Instead, as outlined above, they have to be seen as predictors for future development of financial benefits.

3.2.4 Consequences for Comparable Company Valuation with Special Regard to the Choice of the Valuation Model

Some general remarks regarding the selection of the valuation model can be made based on the above discussions about the dimension of market value to be determined (equity or firm value), the proper use of certain bases of reference and the need for combination models. All statements are based on the assumption that there are no factors other than the value relevance of the bases of reference, the future similarity between the target company and the comparable companies, as well as the technical limitations of some valuation models that might influence the selection process:

125 See DeFusco et al. (2001: 397). The assessment of the regression fit is usually done by an analysis of variance, see Greene (1997: 84, 250-257).

126 Regarding the assumptions underlying OLS regression models (of which most pertain to the model's error term), see DeFusco et al. (2001: 432).

127 See sections 4.3.2.2 and 4.3.2.3.

128 See Herrmann (2002: 121). Regarding the value additivity principle, see Copeland et al. (2005: 30-31).

a. In most cases of single-factor CCV, it is advisable to favour after-cost bases of reference (when using income statement related bases) because they are assumed to generally have higher value relevance than before-cost bases of reference. Thus EBIT- or net income-based models yield probably more accurate valuation results than models that rely on sales or EBITDA. The reason for this is that the bases mentioned first are more similar to the financial benefits that matter in direct valuation cases (i.e. to cash flows)
   If however, the relation between the target company and the peer group is characterised by temporary differences in COGS or in cost allocation policies – i.e. differences that are expected to even out over time – then sales based models (if the cost allocation policy and/or COGS become similar) or EBITDA based models (if the cost allocation policy becomes similar) might indeed lead to more accurate valuation results. Similarly, if after-cost bases are negative while before-cost bases are positive figures then appraisers have to rely on the before-cost bases anyway.
b. Sometimes it is argued that enterprise value bases of reference are superior because they allow for the valuation of companies with different capital structures (see Küting and Eidel, 1999: 229; Benninga and Sarig, 1997: 324-325; Krolle, 2005: 47). This, however, is not totally correct. In general, two companies with noticeably different capital structures cannot be seen as substitutes – no matter what basis of reference is used in CCV (see also Coenenberg and Schultze, 2002: 700-702). The main reason for this is that different portions of debt financing lead to a different financial risk and therefore to a different cost of equity. As a consequence, companies with different capital structures and all other things being equal have to be valued differently (see Herrmann, 2002: 80).
   If, however, these differences in the capital structure are only temporary then enterprise valuation models might indeed be superior to equity valuation models because the bases of reference are not affected by this transitory disequilibrium.
c. The price/sales ratio is not an appropriate single-factor valuation model. This is the case because the numerator and denominator in single-factor multiples should usually correspond with respect to the capital claims, i.e. both figures should either belong to equity holders or both figures should belong to equity and debt holders. As regards the price/sales ratio, however, the numerator (price) is a pure equity figure while the denominator (sales) is calculated before considering financing activities and therefore belongs to all capital providers. This mismatch gives rise to the fact that the price/sales ratio is not a reasonable valuation model from an economic point of view.
d. The use of free cash flow as a basis of reference is not appropriate. This is the case because – if used as a static performance benchmark – it is not possible to find out whether a low free cash flow is a bad sign (the result of operating weakness) or a rather good sign (the result of high capital investments) and vice versa. This, in turn, means that companies would be valued low just because they invest heavily in fixed assets and working capital. This is an obvious paradoxon and, thus, the use of free cash flow cannot be legitimated from an economic point of view.

e. High-growth companies and temporarily underperforming companies often lack positive net income (and possibly also EBIT and EBITDA) figures. As a consequence, when performing classical single-factor CCV, net income (and possibly also EBIT and EBITDA) is not applicable as basis of reference. Thus, because of technical restrictions, appraisers often have to rely on bases of reference that are of minor value relevance, which might be a major problem in CCV.
   Appraisers have to assess case-by-case whether the use of a low value relevance valuation model still allows performing reasonable business valuation. In any case, the uncritical application of e.g. sales based models – as has been done sometimes during the high tech boom in the late 1990s – is not consistent with valuation theory and might often result in dramatic mis-valuations.
f. Single-factor models are not reasonably applicable for negative bases of reference due to technical restrictions. However, appraisers are also advised not to apply these models for zero value or low positive reference variables. While technically applicable, classical single-factor models lack any economic foundation in such a case. In fact, single-factor models cannot assign reasonable values to these companies, because they do not consider that real 'value-reference variable' relations rarely go through the origin (e.g. in reality "zero earnings" does not automatically mean "zero value"). Consequently the valuation results are often biased for "few-cent stocks" (i.e. there may be PE ratios of up to 10,000) (see Damodaran, 2002: 459). Nevertheless, it remains upon the assessment of the appraiser for which minimum value of the reference variable single-factor models yield economicly sound valuation results.
g. Valuation theory predicts that there might be multi-factor models that have higher value relevance than single-factor models. For example, following the ideas of Ohlson (see Ohlson, 1995), models that include current book value and earnings should be able to explain most of the variation in market values. However, it remains to be shown how each variable should be weighted and how these models should effectively be structured so that superior valuation accuracy could be expected.

3.2.5 Implications for the Selection of Comparable Companies

The discussions about the choice of the valuation model have a direct impact on the other main part of CCV: the peer group selection. Two points are worth mentioning: (1) Each valuation model has its own similarity requirements and (2) some models explicitly exclude certain companies from the peer group.

(1) As highlighted in section 3.1.1, the valuation-theory driven approach to peer group selection prescribes that every single-factor model requires the comparable companies to have certain unique similarity characteristics. As a consequence, different models typically require a different set of comparable companies. Even if appraisers rely on the practice-oriented selection approach, these differences remain. The clear focus on industry classification in this approach aims at identifying companies with identical financial benefits' growth rates. In-

dustry classification is, however, not a substitute for other model unique criteria (e.g. similar capital structure when applying the EV/EBIT ratio, or similar net margin when applying the EV/sales ratio). Nevertheless, in real valuation cases appraisers often do not consider this and use the same peer group for every single-factor model.

Multi-factor models aim at explicitly including multiple value drivers in the valuation model and therefore at reducing the similarity requirements for the set of comparable companies. Thus, the number of companies that could principally serve as comparable companies should be higher in the case of multi-factor models than in the case of single-factor models. The determination of the similarity requirements, however, is not as straightforward as it is for single-factor models. Anyhow, a valuation theory derived criteria index for a multi-factor model that includes book value of equity and net income can be found in section 4.3.2.4.

Fig. 17. Impact of valuation model choice on the number of comparable companies available

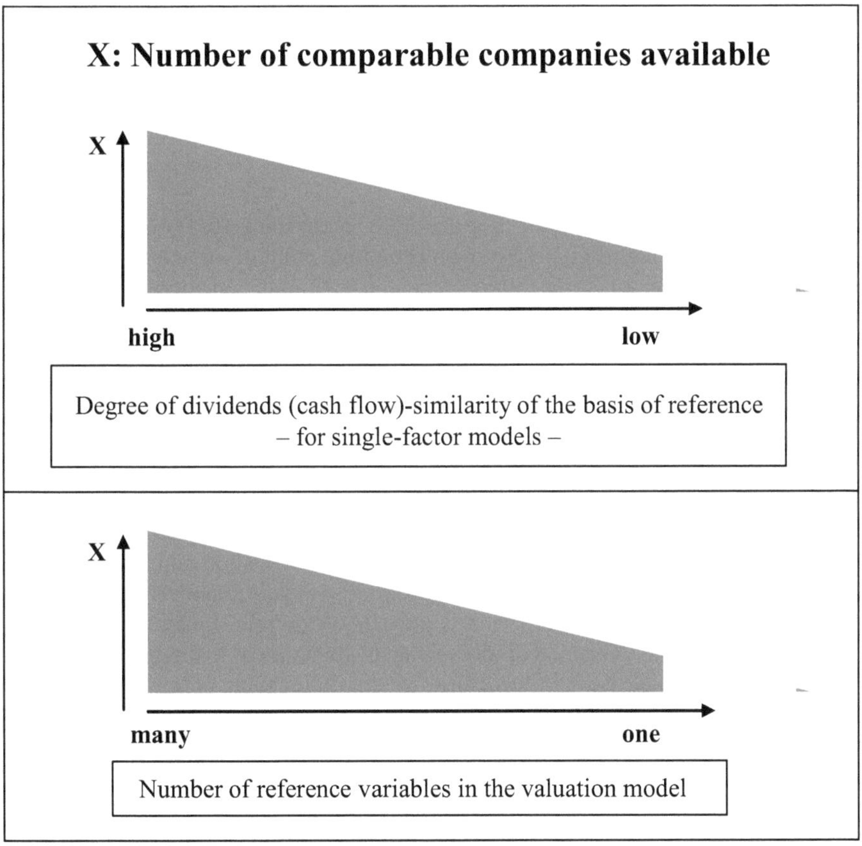

Source: ZEW

(2) Using classical single-factor models – i.e. applying some sort of simple average aggregation approach – definitely sets limits to the selection of comparable companies. These models typically do not allow for the inclusion of companies with negative reference variables and, thus, the peer group typically consists only of companies with positive reference variables.[129] Beyond these purely technical restrictions, appraisers are additionally well advised not to include companies with low positive bases of reference because of economic reasons.

Figure 17 emphasizes the impact of the valuation model choice and the comparable company selection. For single-factor models, the similarity requirements are much lower the more similar the reference variable is to cash flows – a conclusion that directly comes from section 3.1.1. This, in turn, means that there are principally more comparable companies available for cash flow affined bases of reference (in the best case: cash flows themselves) than for those that are very different from cash flows (i.e. sales). Moreover, multi-factor models aim at loosening the similarity requirements for the comparable companies. They therefore allow a principally higher number of companies to serve as comparable companies.

3.3 Comparable Company Valuation as an Integrated Process

Lessons learned

The lesson learned from chapter 3 is that CCV is an integrated process in which all the parts should interdigitate. From an economic point of view it is not possible to determine the set of comparable companies without considering the chosen valuation model. Likewise, appraisers cannot reasonably choose valuation models without accounting for the availability of comparable companies. This availability, in turn, is a function of the similarity of companies and the degree of market efficiency/pricing quality. Figure 18 highlights this integrative approach.

The "solution package"

A sound start to CCV is an "inventory taking"; i.e. to determine which companies can serve as part of the basic population of comparable companies. Such companies can be characterised first by similar industry classification and similar capital structure. After that, a set of generally appropriate valuation models should be determined. The main decision criterion is now the value relevance of the reference variables but also the existence of any technical limitations. If the target company is expected to change its cost structure, its cost allocation policy or its capital structure during the next few years, then the discussions about the future similarity

129 As a solution to this problem, it is sometimes proposed to aggregate the market values and the bases of reference for all comparable companies and to calculate the aggregate multiple afterwards, see Damodaran (2002: 459). This approach however, does not enjoy widespread popularity in valuation practice. Additionally, it is questionable whether this is consistent with valuation theory.

of the target company and the set of comparable companies should also be accounted for.

In the next step, these models should be reviewed and checked as to whether they specifically allow for a sufficiently large set of comparable companies. This should be done considering the discussions concerning the similarity criteria with respect to comparable companies and the degree of pricing quality. Appraisers possibly also have to account for any technical restrictions. In the last step, the model that best fits the requirements and that promises the most accurate valuation results should be performed. If more than one model is principally appropriate then all of them should be applied because of the need for plurality of valuation approaches. Contrarily, if the appraiser feels that none of the models would yield reasonable valuation results then CCV in general is not an apt valuation approach in this case.

It is worth noting that in most cases the efforts to be done in the whole valuation process are absolutely comparable to the efforts necessary to perform a DCF-valuation.[130]

Fig. 18. The integrated process of comparable company valuation

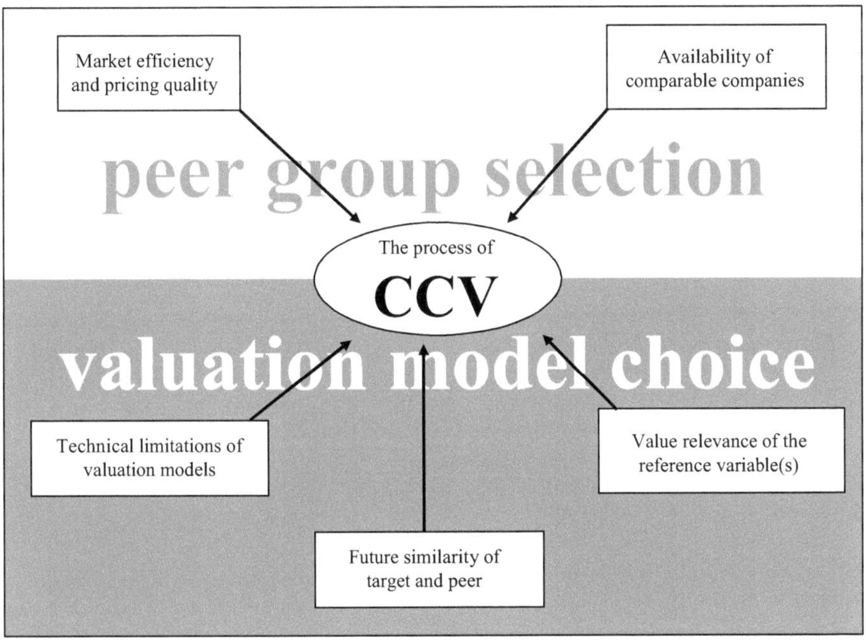

Source: ZEW

130 This opinion is also shared by Moser and Auge-Dickhut (2003: 222).

4 Processing Comparable Company Valuation

This section deals with how to process the different variants of CCV: immediate, single-factor and multi-factor models. The focus of the outlines is clearly on multi-factor models for two reasons. First, immediate CCV models are theoretically not challenging and do not play any role in valuation practice. Only a short explanation about immediate CCV is presented in section 4.1 in order to put across a complete picture of CCV but no in-depth discussion about these models is necessary here. Second, because single-factor models are by far the most dominating models in practice, a lot has been written about them. Literature is full of descriptions about how to process classical models, such as the PE ratio or the PB ratio. Therefore, section 4.2 is restricted to a short summary of the most important issues of the single-factor CCV process. Additionally, some common misinterpretations and practical problems associated with these models are presented in detail. This description of theoretical and practical challenges associated with single-factor CCV directly leads to the presentation of multi-factor models. In section 4.3 an overview of some existing multi-factor models as well as the theoretical derivation of a new multi-factor model based on book value and earnings is given.

4.1 Immediate Comparable Company Valuation

4.1.1 Valuation Process

Immediate CCV models are based on the observation that – in an efficient market – shares of identical companies should have the same price. In this context, "identical" means that the target company and the comparable companies should exhibit the same value-driving factors (financial benefits *FB*, cost of capital *c* and growth rate of financial benefits *g*):[131]

$$V_{it} = V_{jt}$$

$$\text{if } FB_{i\tau} = FB_{j\tau} \text{ and } c_i = c_j \text{ and } g_i = g_j$$

[131] See section 2.3.1 for a more detailed description of the theoretical foundations of immediate CCV.

To process immediate valuation in an efficient market, appraisers have to find a set of companies that is on average identical to the target company (or a set of companies of which every single company is identical to the target company). In such a setting, the price of the target company equals the average price of the set of comparable companies.

Of course, markets have to be sufficiently efficient – or existing inefficiencies have to be adequately accounted for by appraiser – in order to receive accurate appraisal values.

4.1.2 Problems Associated with Immediate Comparable Company Valuation Models

If the requirements are fulfilled, immediate CCV is a very straightforward approach to value a company. However, while immediate CCV can be seen as the "mother" of all CCV approaches because of its close connection to the principle of arbitrage, it is associated with a major shortcoming: Companies are in no way standardised products and – as a consequence – it is difficult to find comparables because of the limited number of companies principally available.

This becomes clear when looking at the German market.[132] The basic population of potential comparable companies does not include all of German companies but only the number of listed German companies. This is the case because only listed German firms have stock prices readily available. In July 2003 there were 987 German companies quoted at German exchanges. To ensure a sufficiently high degree of liquidity and therefore market efficiency, however, appraisers can only focus on companies quoted in the two major segments of the German market: the Prime Standard and the General Standard. This reduces the number of companies principally available for CCV to 693 as of end of July 2003 (see DAI, 2003: 15).

Bearing in mind this small number of companies it becomes obvious that it is very unlikely – even if principally possible – to find two or more companies that are identical in the three main value drivers.[133] Moreover, even if there are identical companies it might be difficult to recognize their identity. Of course, the identity in financial benefits and cost of capital is easy to discover. However, in many cases it is not easily possible to assess whether companies have equal *growth rates* of financial benefits. The problems associated with determining the growth rates of financial benefits are thoroughly addressed in sections 4.2.2.4 and 4.2.2.5. It is,

---

[132] It is assumed here that companies of different national markets can principally not be regarded as comparable due to different legal, tax and competition related circumstances, see section 3.1.1. In exceptional cases, foreign companies that are listed at domestic exchanges can be seen as sufficiently integrated into the domestic market. Moreover, in valuation practice the set of comparable companies often consists of international companies due to small national peer groups.

[133] Even on the US market, with many more companies quoted) immediate CCV models are not relevant at all in valuation practice.

however, important to note that these problems of discovering the equality are of minor importance in practice, compared to the problems discussed above regarding the non-availability of comparable companies.

To sum up: It has to be stated that – mainly due to the non-existence of a sufficiently high number of comparable companies – immediate CCV approaches share almost no popularity at all on capital markets and are of no importance in business valuation. Nevertheless, the underlying concept (principle of arbitrage) is overwhelmingly straightforward. Therefore, these approaches have to be understood as the methodological basis of (and pave the way for) the more sophisticated single-factor and multi-factor CCV models.

4.2 Single-Factor Comparable Company Valuation

4.2.1 Valuation Process

Single-factor CCV is by far the most relevant CCV approach in practice. Many of the points that are important in this section have already been discussed in great depth in chapters 2 and 3. Thus, on the one hand the description below can be understood as a wrap on the main issues that are especially essential in *single-factor* CCV. On the other hand, however, this section also provides a compact guideline on how to proceed with single-factor valuation and on how to avoid common mistakes. While it has been outlined in section 3.3 that CCV should always follow an integrated process of selecting the basis/bases of reference and selecting the comparable companies, the single steps are presented separately below for a better understanding.

Choice of the basis of reference

In the first step appraisers have to assess whether the model chosen has any technical restrictions. If the appraiser decided to use – as in most cases – classical single-factor models that rely on a simple average aggregation mechanism (such as the PE ratio or the PB ratio), reference variables cannot be negative. In this context it is important to note that accounting figures that stand at the top of the income statement have a higher probability of being non-negative than those that stand at the bottom of the income statement; i.e. in most cases sales is bigger than EBITDA which is bigger than EBIT which is bigger than net income. If the appraiser decides to perform a regression-based aggregation of peer group data, however, there are no restrictions concerning the choice of the basis of reference.

From the resulting set of technically applicable bases of reference the appraiser has to choose the one that has the highest expected value relevance. To determine the degree of value relevance the theoretical discussions in section 3.2.1 might be helpful. An important constraint to this value relevance assessment is the future similarity of the target company and the comparable companies. If companies presumably change their cost structure or capital structure in the future then other than the most value relevant reference variables might be advisable. section 3.2.2 gives guidance on how to account for this issue.

It is worth noting that the use of enterprise bases of reference (bases that are calculated before considering the financing of the company, such as sales, EBITDA, EBIT, or total assets) requires building multiples that relate the variable to the value of the whole firm (the value of both equity and debt; the so-called enterprise value). Contrary to that, the use of equity bases of reference (bases that are calculated after considering the financing of the company, such as net income or book value of equity) require building multiples that relate the variable to the stock price (i.e. the market value of equity).

Two common mistakes when choosing the bases of reference are (1) the use of free-cash flow bases in general (as proposed e.g. by Damodaran, 1996: 309-312) and (2) the combination of price and sales when putting together the multiple (as proposed e.g. by Damodaran, 2002: 544-551; Stowe et al., 2002: 216-222; see also section 3.2.1.2). (1) The problem in the first case is that the reference variable is negatively affected by two conflicting components: (value diminishing) costs and (typically value enhancing) investments. To put it more clearly, current free cash flow might be low due to high costs – which is reasonable – or due to high investments – which is unreasonable. The negative effect of investments is not in line with the central underlying principle of single-factor CCV which says that the reference variable and the value of the company should be non-ambiguously positively related. Therefore, current free cash flow is a very bad indicator of future free cash flow and should not be applied as a basis of reference.[134] (2) In the second case, appraisers compare a pure equity figure (i.e. the stock price) with an enterprise figure (sales). The financial inequality of numerator and denominator results in biased valuation results, especially at levels of high debt financing of the companies that are involved in the valuation process.

Selection of comparable companies

First, appraisers have to identify fundamentally similar companies. This similarity heavily depends on the applied valuation model. By rearranging the Gordon growth model the value driver for every single-factor CCV model can be exposed.[135] A company qualifies for the basic set of comparable companies if it fulfils all similarity criteria; the long-term growth rate of financial benefits is always one of these criteria.

The identity in many similarity criteria – such as the cost of capital or the gross margin – is more or less straightforward to detect, because the respective accounting figures are currently available to or can be calculated by appraisers. Contrary

---

[134] See also section 3.2.1.4. Clearly, if appraisers correctly account for the growth rate of free cash flows then CCV is also possible using these cash flows as a basis of reference. However, while it is very difficult to determine the growth rate of financial benefits in general (see section 4.2.2.5) it is almost impossible to determine the growth rate of free cash flows of a company that is not in equilibrium, see Bamberger (1999: 658-660).

[135] See section 3.1.1 for more details.

to that, to find companies with similar long-term growth rates is not quite so straightforward. To manage that task, appraisers are typically advised to use the industry classification as a proxy for identical future growth. However, that is not enough because the growth rate also depends on other determinants.[136]

After having established the basic set of comparable companies, the selection should continue with a closer look at capital market efficiency and pricing quality. As in the case of immediate CCV, it is also important to note here that even in a perfectly efficient market not every company is priced correctly but they are on average priced correctly (see Damodaran, 2002: 113). This leads to the recommendation to always put together a set of comparable companies (to balance minor mispricing of single companies) rather than to use a single comparable company. Since the German market is characterised by a generally high degree of – but in no way perfect strong form – efficiency it is additionally advisable to analyse the set of comparable companies as to whether they are systematically mispriced. Indications for such a systematic mispricing are e.g. a recent IPO or the involvement in a takeover bid but also size differences or lack of covering by financial analysts. If quoted companies with a low degree of liquidity (companies that are not frequently traded) are part of the valuation process then even more care has to be taken when assessing the pricing quality. Companies that are assumed to be systematically mispriced compared to other companies have to be eliminated from the peer group.

Two common mistakes when selecting comparable companies are (1) the incorrect handling of the long-term growth rate[137] and (2) ignoring the difference in similarity requirements for each valuation model (see e.g. Barthel, 1996: 150; Cheridito and Hadewicz, 2001: 322). Both mistakes are closely connected. (1) In the first case, appraisers are aware that long-term growth is a comparability criterion. However, they do not proxy it correctly. In most cases, it is simply the industry classification that determines the peer group, neglecting that there are other important influences on growth. (2) In the second case, appraisers use a unique set of comparable companies for several single-factor valuation models, such as the PE ratio and the EV/sales ratio. Therewith, they ignore that each model has different requirements to the selection of peer groups.

In contrast to the two common mistakes that were presented above (as part of the sub-section "choice of the basis of reference"), the two mistakes discussed here do not violate the valuation principles. They are rather the result of a bad handling and interpretation of similarity requirements. Of course, it is worth noting that in CCV complexity reduction is necessary – especially when dealing with growth rates. However, too much of it is contra-productive. Refer to sections

[136] An in-depth discussion about the determinants of long-term growth of financial benefits and the problems associated with accurately accounting for this growth rate in single factor models is presented in section 4.2.2.5.

[137] Sometimes it is even unknown that the growth rate actually is a fundamental similarity requirement: "For example, application of a price earnings multiple does not require explicit specification of a firm's cost of capital or growth rate." Palepu et al. (1996: chapter 7, 16).

4.2.2.4 and 4.2.2.5 for details on how to deal with the factor "financial benefits' growth".

Aggregation process
Typically, appraisers aggregate the peer group results by using one of the simple averaging variants (arithmetic mean, harmonic mean, median) described in section 2.4.2. If the requirements are fulfilled, the regression approach can alternatively be applied. As compared to the simple average aggregation methods, the advantage of this approach is that in most cases it assigns a positive corporate value to the zero-value or low negative bases of reference. This is rather in-line with reality at capital markets.

Outliers (companies with extreme multiples) are usually eliminated from the peer group (see Damodaran, 2001a: 260).

4.2.2 Problems Associated with Single-Factor Comparable Company Valuation Models

4.2.2.1 Non-Negativity Restriction of the Bases of Reference

A major problem when applying classical single-factor CCV (i.e. using the simple averaging aggregation methods arithmetic mean, harmonic mean or median) is that no reasonable (positive) corporate value can be determined if the basis of reference is negative. This is due to the calculation mechanisms which – for a given multiple $\left[V_{jt} \cdot \left(BR_{j\tau}\right)^{-1}\right]$ – bring forth a straight valuation line that passes through the origin of a two dimensional "value – basis of reference BR_i"-coordinate system:

$$V_{it} = \left[V_{jt} \cdot \left(BR_{j\tau}\right)^{-1}\right] \cdot BR_{i\tau}$$

As a result, positive bases of reference of the target company ($BR_i > 0$) yield positive corporate values, zero value bases of reference ($BR_i = 0$) yield a zero corporate value and negative bases of reference ($BR_i < 0$) yield a negative corporate value (assuming the multiple to be positive).[138] At real capital markets, however, one can find many companies with contemporaneous negative current accounting variables and positive stock prices. Figure 19 shows how some of the most commonly used bases of reference relate to the stock price or the enterprise value respectively. Additionally, the multiple line (i.e. the line that would result in single-factor CCV when using the arithmetic mean as an aggregation method) is plotted in the graphs.

---

[138] See also section 2.4.2 and 3.2.3.

The conclusion to be drawn from this figure is that the non-negativity restriction for the reference variables is of special importance in terms of net income and operating cash flow followed by EBIT and EBITDA. Applying sales and book value as reference variables generally yields no major problems because these figures are positive in almost every case. This finding is in line with the theoretical analysis in section 3.2.3.

Based on this finding one might presume that classical single-factor models cannot reflect the reality at capital markets very well. But what are the consequences? If appraisers face negative reference variables in practice they just switch to other (positive) bases and use the latter ones in their valuation models. This might lead to a situation where appraisers are forced to apply low value relevance figures just because these are the only variables that are applicable from a technical point of view. The typical result of such behaviour is a poor valuation quality. However, appraisers should be aware that not only negative bases are associated with problems. The application of low positive bases of reference in classical single-factor CCV models – although principally possible from a technical point of view – often leads to the same poor valuation quality.

Fig. 19. "Value – reference variables" associations

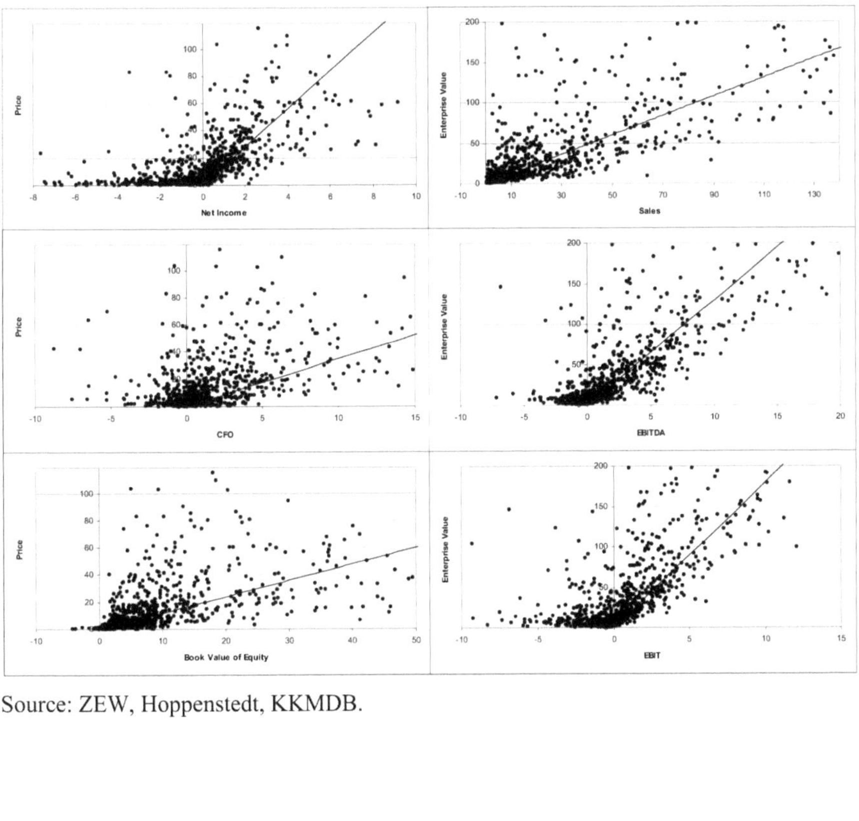

Source: ZEW, Hoppenstedt, KKMDB.

4.2.2.2 High Requirements of Comparable Companies

The valuation theory-driven approach to assessing the similarity of companies (section 3.1.1.) has relatively high requirements to comparable companies in terms of financial characteristics. Based on the Gordon growth model it could be shown that e.g. the PE ratio requires the comparable companies to exhibit the same cost of equity, payout ratio and growth rate of dividends. Other classical single-factor models have even stricter requirements. Of course, the requirements are much lower than in the case of immediate CCV, but they are still tight.

This would not be a problem in a world with an infinite number of listed companies. However, at real stock markets this is not the case. As has been outlined in the discussion about immediate CCV, companies that principally qualify for CCV on the German market are scarce. Among these few companies appraisers now have to select a sufficiently large set with similar comparability criteria. It is needless to say that the number of companies that agree with the relevant characteristics is typically very small. There might be companies with (coincidentally) several comparables but there might also be companies with very few or even no economicly sound comparables. The consequence for CCV is that in many cases it is not easy – sometimes even impossible – to compile a sufficiently large set of comparable companies that can reasonably be justified by valuation theory.

To sum up: The requirements of comparables are much lower than in the case of immediate CCV, but they are still high enough to cause a shipwreck of the valuation process in some cases.

4.2.2.3 Is Value Relevance an Issue?

There might be indications that in some cases single accounting figures have low value relevance. For example, for a company that is characterised by an expected semi-strong persistence of abnormal earnings both book value and earnings have joint value relevance, but neither of these two variables have high individual value relevance.[139] However, it remains to be shown in the empirical part of this study (section 5.2) whether – and if yes: Under which circumstances – this is really a problem in the context of CCV.

4.2.2.4 Dealing with the Long-Term Growth Rate

The similarity determining factor that is the least predictable is the long-term growth rate of financial benefits. Other similarity determinants are usually available at the date of valuation. In contrast, the long-term growth rate is not; instead, it requires a multi-period forecast, which is typically associated with a huge amount of uncertainty.

---

139 See section 3.2.1.3.

Fig. 20. The difficulty of determining the long-term growth rate (in the eyes of analysts and investors)

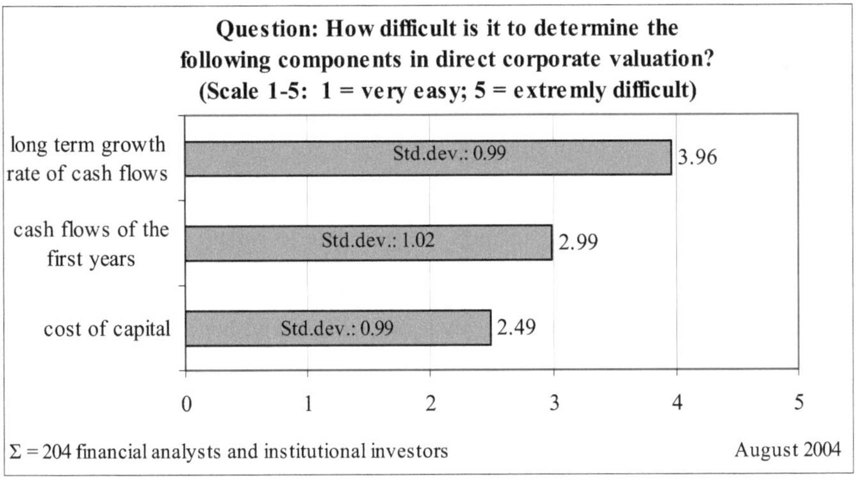

Source: ZEW

Fig. 21. Effect of the growth rate on the corporate value

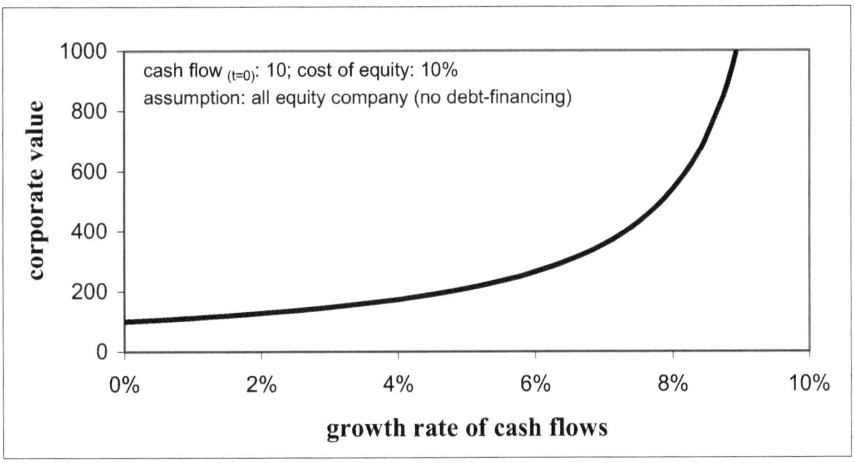

Source: ZEW

A survey among 204 financial analysts and institutional investors, conducted as part of the ZEW financial market survey in 2004 (see Meitner, 2005: 8), confirms this. The financial experts comment that the determination of the long-term financial benefits' growth rate is by far the most difficult job when applying direct company valuation models. Figure 20 shows that determining the cash flows in the detailed forecasting period as well as the cost of capital is sensed as being much easier.

The forecasting problem is further complicated by the fact that corporate values react very sensibly to a change in the growth rate. Figure 21 helps to clarify the relation between the long-term growth rate and the corporate value. To preserve ease of illustration this is done by using the Gordon growth model.

This all makes the long-term growth rate one of the most delicate issues in business valuation. The points that are particularly worth thinking about when performing CCV are the following: Is it possible to accurately determine the long-term growth rate within a reasonable time for the target as well as for the comparable companies? If yes: How? If not: Is it possible to isolate variables based on which long-term growth can roughly be estimated? To answer all these questions, a closer look at this issue is necessary. Therefore, the next section sheds some light on factors that influence the growth rate of financial benefits and presents a way to determine the growth rate step-by-step.

4.2.2.5 Excursus: How to Determine the Long-Term Growth Rate of Cash Flows or Earnings[140]

In contemporary literature the issue of "long-term growth rates in the context of business valuation" has failed to attract the intensity of scientific research that other areas of valuation theory have accomplished.[141] Part of the explanation can be found in the fact that a fair amount of uncertainty exists when forecasting future events, which makes it difficult to admonish specific actions. Additionally, it is the interaction of particular accounting issues, corporate strategies, and economic variables, which as an entity is difficult to be described.

Critical assessment of some suggestions from literature

There exists an exhaustive amount of material suggesting numerous ways to determine long-term financial benefits' growth rates. The most important suggestions are presented below.

- Linear or non-linear extrapolation either by means of regression or by simple long-term averaging (see e.g. Hail and Meyer, 2002: 573-584; Damodaran, 2001a: 149). Extrapolation of preceding growth rates is not advisable because this would violate the principle of future orientation of valuation (see Moxter, 1983: 97-101; IDW, 2002: 59). Indeed, this principle allows a recursive data analysis to improve the forecasting quality, but – due to the timely impermanence of the company itself and of its environment – historical values cannot be brought forward without a deeper analysis of the specific valuation situation. In this context, recent research shows that this method

[140] The following outlines are mainly based on Meitner (2005).

[141] Recent exceptions are provided by Albrecht (2004); Chan et al. (2003) and Henselmann (2000).

also has to be rejected from a statistical point of view (see Chan et al., 2003: 655).

- Using consensus forecasts for listed companies (see Hail and Meyer, 2002: 584; Damodaran, 1997: 624). While short-term forecasts are on average correct, long-term estimations result in progressively deteriorating accuracy (see Chan et al., 2003: 683; Damodaran, 2001a: 157-158). In addition, it is very rare for analysts to provide projections which exceed five years and, thus, their estimations only cover parts of the long-term growth rate. Hence, their use as a substitute for the long-term growth rate seems to be very limited.[142]
- Economic growth as a benchmark (see Baetge et al., 2005: 338; Schultze, 2003: 74; Copeland et al., 2000: 279). Suggestions for long-term growth rates of financial benefits range from using gross domestic product (GDP) growth rates (both real and nominal) to the growth rate of private consumption and even as far as the inflation rate. However, a restricted focus on only macro-economic indicators for determining the growth rate is too narrow. Indeed, this method can be used to describe the growth rate of sales of a company if that company is in state of equilibrium[143] (which might be several years in the future from the date of valuation), but would be insufficient to describe the long-term growth rate of financial benefits starting at the date of valuation. The following section is dedicated to exemplify this.

Growth determining factors from an economic perspective

It is undisputed that macro-economic trends have an effect on company growth. GDP growth puts a *long-term* upper boundary on company growth because otherwise the volume of a company would exceed that of the country (see Schultze, 2003: 74). However, it is crucial to make clear what is meant by company growth in this context. GDP is defined as the value of all goods and services produced in an economy within a particular time span (see e.g. Blanchard, 1997: 20) or – to put it differently – GDP is nothing more than an indicator for the revenues or sales generated in an economy. Consequently, this GDP restriction only holds for the growth rate of *sales* of a company.

However, GDP growth is not the only source of influence on the growth rate of sales. Regarding the *medium-term* development of the company, specific industry variables also have a noteworthy effect. For example, the degree of competition in an industry crucially determines the time span in which a company can generate sales numbers that are above average and how quickly the sales growth rate ad-

[142] A contrary opinion is provided by Herrmann (2002 who found that consensus forecasts for earnings have at least partial explanation power for the long-term growth rate of earnings, see Herrmann (2002: 234).

[143] In direct valuation approaches, such as the DCF-method or the earnings capitalisation variants, it is appreciable to perform a detailed forecast of cash flows until companies reach the steady state. Once this date is reached, the Gordon Growth Model can be applied to determine the terminal value, see Peemöller and Kunowski (2005: 230). CCV, however, requires the growth rate as from the date of valuation.

justs to its long-term value (see e.g. Porter, 1980: 142-145). Contrarily, company-specific factors have only a limited influence on sales growth.[144]

Modeling the growth rate of financial benefits

The preceding analysis provided a useful basis for establishing the growth rate. Nevertheless, a closer look at the characteristics of the relevant growth rate in the context of CCV is necessary. Two aspects are of particular importance: (1) The long-term growth rate that is relevant in CCV is the average growth rate beginning at the date of valuation, and *not* the growth rate that a company reaches in the state of equilibrium. (2) In CCV, growth rates of financial benefits are relevant and *not* growth rates of sales.

As discussed above, the first of these two issues implies that the long-term growth rate is not only comprised of country-wide economic factors but also of industry-specific influences. These industry-specific influences have an impact on the company's sales especially in the first few years after the valuation date. The conclusion to be drawn from the second aspect is that – in order to determine the growth rate of financial benefits – sales at the date of valuation as well as the sales-cost situation of the company need to be taken into consideration. This, in turn, means that (in contrast to the determination of the growth rate of sales) company-specific factors are of significant importance.

The determination of the growth rate of financial benefits follows two steps:

(1) It is necessary to determine the growth rate of sales at the time of valuation. This can be done by using one of the valuation approaches discussed below.

- Long-term economic growth
  In the simplest case the company is in a steady state. Thus, the sales growth rate is expected to be stable and can therefore be approximated using macro-factors. However, it must be remembered that economic growth is only an upper boundary. A sales growth which lies under the GDP growth rate is anything but implausible.
- Phase Models
  In practice it is common that companies are not in a state of equilibrium at the date of valuation. In most cases, companies reveal a higher sales growth rate during the first few years after the valuation than during later stages. This cognition led to the development of models that specifically account for the possibility of a dynamic growth rate. One of them stems from direct valuation models. In the so-called 2-phase model, an interim growth rate is applied (phase 1) until the stable long-term growth rate becomes effective (phase 2) (see Stowe et al., 2002: 72-73). These two growth rates can be "aggregated" to determine the average growth rate. From a technical point of view it is important to note that it would also be possible to apply more than just two phases.

144 There should only be some short- and medium-term effects that are due to a highly competent management.

- The H-Model
  The H-Model assumes a high growth rate at the beginning of the general planning phase, which declines linearly over time and eventually converges to a (lower) long-term growth rate (see Fuller and Hsia, 1984). In general, a more transparent reflection of reality can be reached than in the 2-phase model because the assumption of a sporadic switchover from a high short-term growth rate to a moderate long-term growth rate is relaxed. In practical applications the H-Model receives reverence in industries where there is only limited competition amongst companies at the time of valuation. When new companies enter the market it should be expected – ceteris paribus – that the growth rate of sales gradually declines until it reaches a stabilising equilibrium (see White et al., 1997: 1101-1102; Stowe et al., 2002: 76-77).
- Fade Factors
  If a very competitive environment causes a degressive reduction in the growth rate, the use of the fade factor model pays to be implemented (see Schwetzler, 2003: 80). In this model, the relation between two consecutive growth rates can be described as follows:

$$g_t = g_{t-1} \cdot (1-\psi) \text{ resp. } g_t = g_1 \cdot (1-\psi)^{t-1}$$

  The fade factor ψ is $\in [0,1]$. A high fade factor results in a faster decline of the growth rate and vice versa. The value of the fade factor can thus be viewed as an indicator for the prevailing competition in that industry.

Fig. 22. Development of the growth rate over time in different valuation models

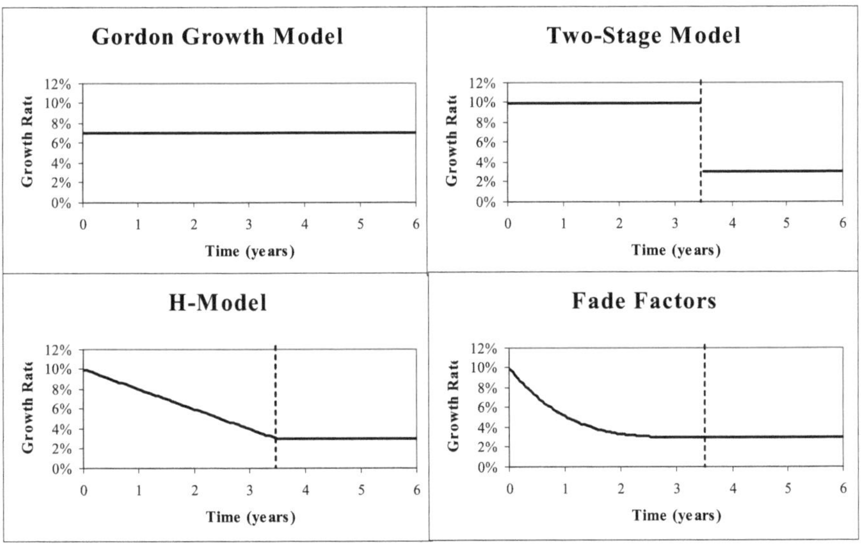

Source: ZEW

- Conclusion
  The long-term growth rate of sales (which is relevant in CCV) can be calculated from all of these models as some sort of "average growth rate" either by means of iteration (see Schwetzler, 2003: 80) or in a closed form calculation. The models represent helpful tools when determining the average growth rate in situations where companies are not in equilibrium at the date of valuation. In summary, Figure 22 graphically depicts the assumed growth development of the models covered thus far.

(2) Once the average growth rate of sales is determined, one can proceed to calculate the financial benefits' growth rate.[145] At this point it needs to be taken into account that a company's total costs are split into fixed and variable costs. Therefore, the relationship between sales and operating profits behaves in the way that is shown in Figure 23.

The figure reveals that while sales is always positive, operating profit is negative at a low sales level. Thus, while the application of a growth rate for sales is possible at all levels, for operating profit it is only reasonable to work with growth rates if sales exceed the threshold a (because of the need for a positive basis of growth). This gives rise to the suggestion that a certain sales growth does not translate into an equal growth rate of operating profits. Consequently, to determine the growth rate of operating profit that – if applied in the Gordon growth model – yields the same corporate value as the application of the already determined growth rate of sales, a simple rearrangement of the equation consisting of two valuation formulas is helpful. The first formula describes the calculation of the enterprise value EV using a constant growth rate of sales g_{SALES}:

$$\begin{aligned} EV_0 &= \lim_{N\to\infty} \sum_{n=1}^{N} \left(SALES_n \cdot (1-v) - F\right) \cdot (1+c)^{-n} \\ &= -F \cdot c^{-1} + \left(SALES_0 \cdot (1-v) \cdot (1+g_{SALES})\right) \cdot \left(c - g_{SALES}\right) \end{aligned} \tag{4.1}$$

The parameter F stands for fixed costs and v measures the proportion of variable costs to sales. The second formula is simply the Gordon growth model on the enterprise level:

$$EV_0 = EBIT_0 \cdot (1+g_{EBIT}) \cdot (c-g_{EBIT})^{-1} \tag{4.2}$$

Equating the right hand expressions of Equations 4.1 and 4.2 and solving for g_{EBIT} yields:

[145] Below, the growth rate of operating profit is calculated exemplarily.

$$g_{EBIT} = \left(c \cdot EV_0 - EBIT_0\right) \cdot \left(EBIT_0 + EV_0\right)^{-1} \tag{4.3}$$

$$\text{with } EV_0 = -F \cdot c^{-1} + \left(SALES_0 \cdot (1-v) \cdot \left(1 + g_{SALES}\right)\right) \cdot \left(c - g_{SALES}\right)$$

Fig. 23. The relation between sales and operating profit

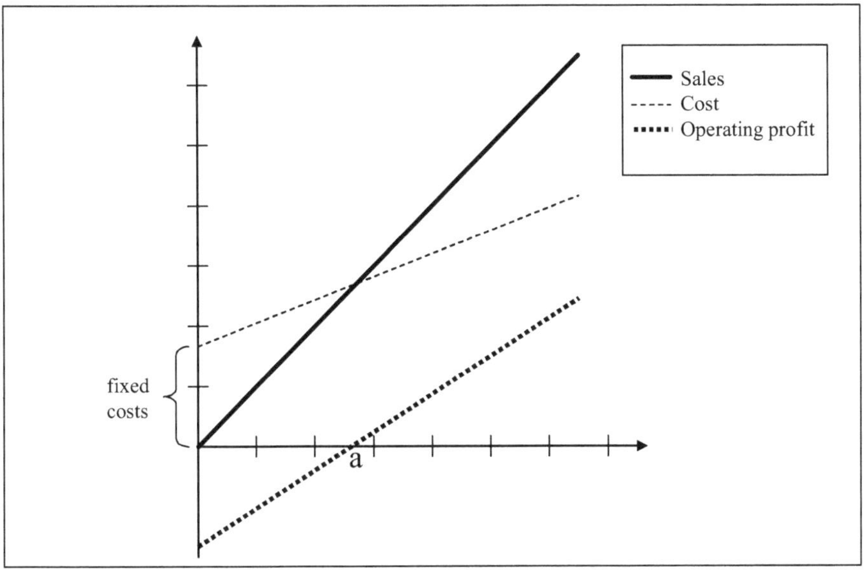

Source: ZEW

Equation 4.3, which is stated for the growth rate of EBIT, is principally valid for all kinds of financial benefits (earnings and cash flows). Differences arise however in the definition of cost. Figure 24 emphasizes the relation between the sales growth rate and the implicit model-based EBIT growth rate. The figure reveals that, especially in the case of a low but positive EBIT, differences arise between the two growth rates at the date of valuation. Thus, special care needs to be taken when determining financial benefits' growth rates for companies, which generate low or even negative profits in the short- and mid-term. However, for companies with a relatively high level of profits at the date of valuation, the difference between the two growth rates decreases.

The results from a ZEW survey (see Meitner, 2005: 10) are largely supportive of the procedure to determine the growth rate of financial benefits explained above. Being asked about the importance of factors influencing the long-term financial benefits' growth rate, experts stated that company-specific characteristics are the most important (none of the participants replied that these factors had no significance whatsoever). Of lesser importance were the industry- and economy-wide influences. According to the analysts, profitability (and therefore implicitly the amount of earnings) contributes most significantly to the height of the financial benefits growth rate. Financing opportunities of companies was given inferior preference.[146] Figures 25 and 26 summarise this.

Fig. 24. The relation between sales growth and implicit EBIT growth

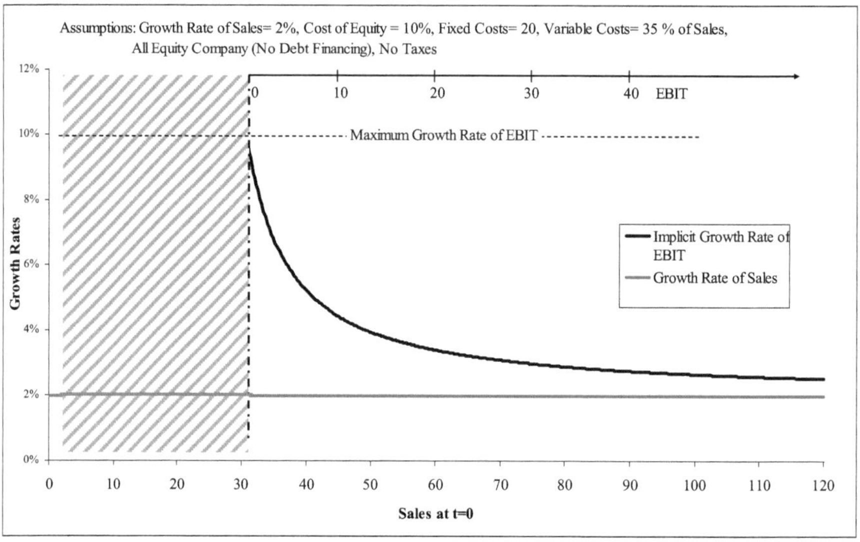

Source: ZEW

To summarise: Three results and conclusions from the last two sections in particular are of importance. First, it has been shown both empirically (via the survey) and theoretically that it is highly difficult to forecast the long-term growth rate. Second, industry classification might be a good proxy for the long-term growth rate of sales, but not necessarily for the long-term growth rate of financial benefits. For purposes of CCV, it is advisable to additionally account for the current profitability or the net margin of the company.[147] Third, to proxy the growth rate by industry plus profitability would further decrease the already small sets of potential comparable companies.

---

146 The set of company-specific influences (profitability, dividend policy, external financing possibilities) is drawn based on Benninga and Sarig (1997: 317-318).

147 Certainly, accounting for industry and profitability cannot totally substitute the growth rate of financial benefits. However, the outlines above showed that it seems to be a good proxy. In this context, the importance of profitability is also supported by recent research: An international study found that in CCV comparable company selection based on profitability outperforms comparables selection based on industry classification, see Dittmann and Weiner (2005).

Fig. 25. Factors influencing the long-term growth rate (in the eyes of analysts and investors)

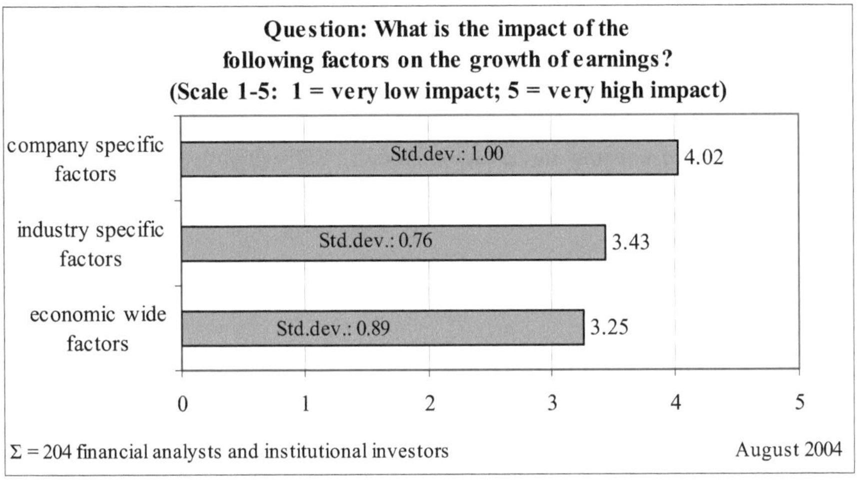

Source: ZEW

Fig. 26. Company-specific variables influencing the long-term growth rate (in the eyes of analysts and investors)

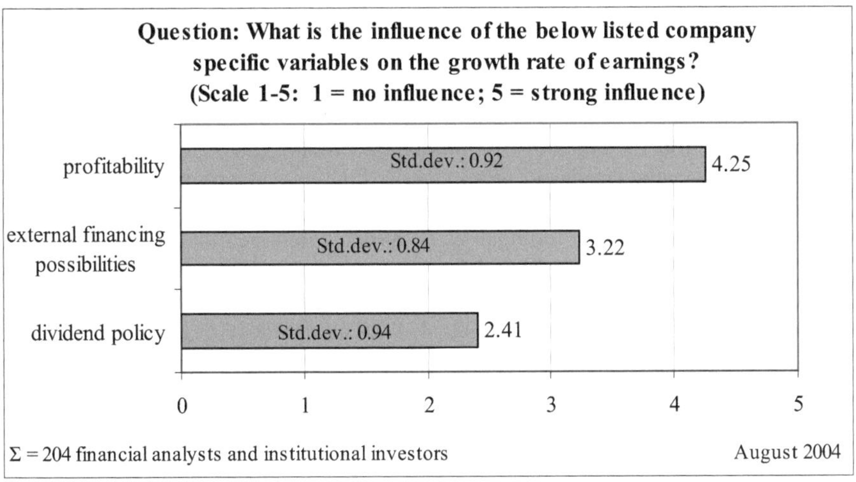

Source: ZEW

4.3 Multi-Factor Comparable Company Valuation

Multi-factor CCV models are characterised by more than one reference variable. Typically, they are more difficult to apply than single-factor models. Therefore, the assessment of such models has to focus on whether they manage to overcome all (or at least some) of the problems associated with single-factor models. The remainder of this section is organised in the following way: It starts with a short overview of some existing multi-factor models. Subsequently, the development and derivation of a two-factor model that is based on earnings and book value is presented.

4.3.1 Existing Models

4.3.1.1 Price-Earnings-to-Growth (PEG) Model

The Price-earnings-to-growth (PEG) model is a CCV approach that relies on two factors: earnings and the expected growth of earnings (see Adrian, 2005: 79; Löhnert and Böckmann, 2005: 413; Wiehle et al., 2004: 69; Schwetzler, 2003: 81-82, Damodaran, 2002: 487-496; Peemöller et al., 2002: 207). It is defined as the PE ratio divided by the respective growth rate. Typically the use of a trailing PE ratio rather than the forward PE ratio is required because otherwise growth would be counted twice (see Damodaran, 2002: 487). The PEG ratio is a valuation model that is often applied in practice. Sometimes it is also used as a simple indicator for over- and undervaluation of stocks. In the latter case, a PEG ratio inferior to 1 means a general undervaluation, while a PEG ratio superior to 1 is a sign of a general overvaluation.

The aim of the PEG ratio is to eliminate the growth factor from the comparable company selection by explicitly including that factor into the valuation model. This would lower the similarity requirements for the peer group and therefore allow for a principally higher number of comparable companies. Remember that a high number of comparable companies is appreciable since this allows balancing unsystematic mispricings or minor differences in the similarity of companies.

A closer look at the PEG ratio, however, sheds light on the fact that this model fails to do so (see Damodaran, 2002: 491). Taking the Gordon growth model (Equation 2.2) as a reference, the comparability criteria for the peer group when using of the PEG ratio are as follows:

$$PRICE_t \cdot \left(NI_t\right)^{-1} \cdot \left(g\right)^{-1} = \left(DIV_t \cdot \left(NI_t\right)^{-1}\right) \cdot \left(1+g^{-1}\right) \cdot \left(c_e - g\right)^{-1} \tag{4.4}$$

Fig. 27. PE ratio and PEG ratio for different growth scenarios

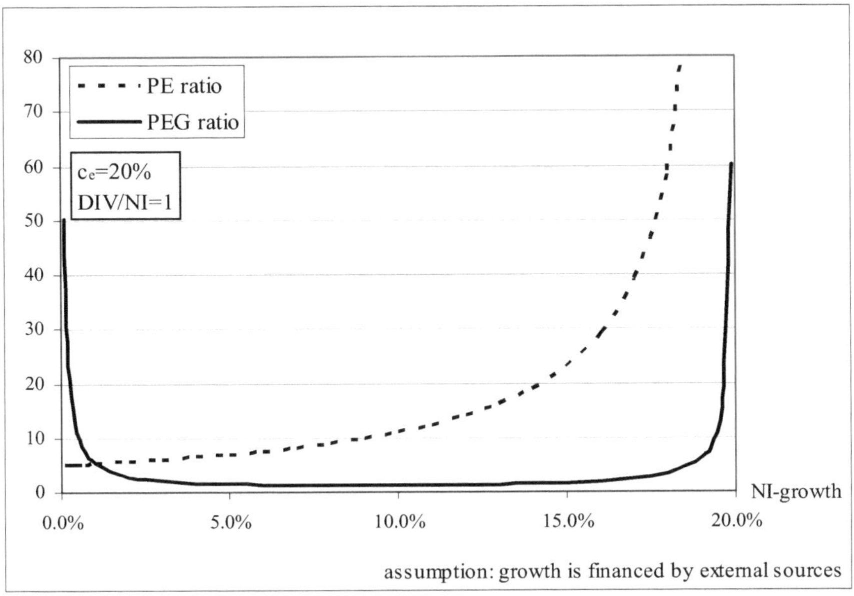

Source: Adrian, 2005: 82, ZEW

Fig. 28. Growth sensitivity of the PE ratio and the PEG ratio

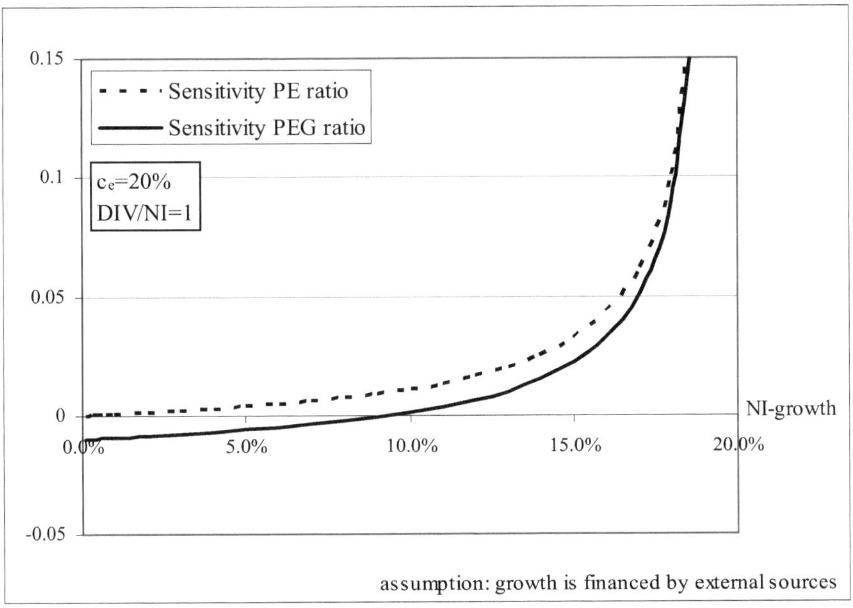

Source: ZEW

Assuming that the clean-surplus relation holds, however, the influence of the payout ratio can be eliminated which leads to the final similarity requirements[148]:

$$PRICE_t \cdot \left(NI_t\right)^{-1} \cdot \left(g\right)^{-1} = f\left(c_e; g\right)$$

with $\Delta PRICE / \Delta c_e < 0$ and $\Delta PRICE / \Delta g \gtrless 0$

While not thoroughly eliminating the influence of the growth rate, the PEG ratio at least succeeds in limiting its influence for certain growth-companies. This is shown in Figure 27 and Figure 28 for a fictitious company with a 20 % cost of equity.

The figures reveal that, while the absolute value of the growth sensitivity of the PE ratio is lowest for no-growth companies and continuously increases with rising growth rates, the absolute value of the growth sensitivity of the PEG ratio has a minimum at a growth rate of about 9.5 % (for the exemplary case). The PEG ratio has lower absolute sensitivities in the range from 6.5 % to almost 20 % and is therefore less affected by the growth rate than the PE ratio. This leads to the conclusion that the use of the PEG ratio is especially advisable for high-growth companies.[149] Contrarily, for low-growth companies the PE ratio is less dependent on the growth rate and is therefore the better choice.

To summarise, it has to be stated that the PEG ratio cannot be seen as a model that is generally superior to the single-factor models for at least two reasons. First, the shifting of the growth rate from the selection process to the valuation model does not prevent appraisers from facing the forecasting problems associated with the growth rate. Second, the intended shifting does not really take place. The PEG ratio only limits the influence of the growth rate in the comparable company selection process but obviously fails to eliminate it. That, in turn, means that the appropriate peer group selection still requires the consideration of the growth rate and, thus, there is no simplification compared to the selection process associated with the PE ratio.

From a technical point of view, it might be advantageous to use the PEG ratio in case the target company is a high growth firm because then this ratio is less sensitive to growth rate changes than the PE ratio. However, it is important to note that the use of the PEG ratio requires the determination of the growth rates of all companies involved in the valuation process. In contrast to that, for classical single-factor models it is only necessary to identify companies with identical growth rates *but not* to exactly determine the growth rates.

---

148 Sometimes it is mentioned that one of the disadvantages of the PEG ratio is that it cannot eliminate the influence of the cost of capital, see e.g. Schwetzler (2003: 82). This elimination, however, is not one of the aims of the PEG ratio; the cost of capital rather has to be considered as a similarity criterion when putting together the set of comparable companies.

149 A similar opinion is provided by Adrian (2005: 81).

4.3.1.2 Multi-Model Approaches

A variant that enjoys widespread popularity in valuation practice is to apply several single-factor CCV models side-by-side and to aggregate the different valuation results in order to finally determine the corporate value. This parallel application of more than one model is principally possible with any combination of single-factor models. The general proceeding is as follows: Usually, appraisers eliminate the outlying results and then construct the corporate value by using a simple equal weighting (or some sort of subjective weighting) of multiples. Unfortunately, there is no theoretical foundation to multi model approaches – with one exception: A justification for a combination of the PE ratio and the PB ratio might be delivered by the Ohlson model.

It is important to note that the Ohlson model (Equation 3.8) does *not* require the inclusion of a growth rate. Contrarily, it requires the determination of certain persistence parameters. The major achievement of the Ohlson model is the demonstration of the formal linkage between value and the two accounting numbers book value and earnings (see Lo and Lys, 2000a; Lundholm, 1995: 761). This becomes obvious when neglecting the information not captured by current accounting. Consequently, a CCV model that relies on both, the PE ratio and the PB ratio, is largely consistent with the basic idea of Ohlson. A general problem is that – while in the Ohlson model the weightings of the two variables are roughly described by the variable k – there is no rule describing how to assign weights to each multiple in the PE/PB combination model.

In literature multi model approaches are sometimes criticised (see e.g. Peemöller et al., 2002: 205). This is due to the two reasons that have been mentioned already: First, up to this point there is no theoretical basis for multi model approaches other than the combination of the PE ratio and the PB ratio. Second, even if a combination model of the PE and the PB ratio is applied, the problem of how to weigh the two multiples remains. However, despite these theoretical concerns, it has been found for the US stock market that even a simple equally weighted PE/PB ratio combination model outperforms both single-factor models: the PE ratio and the PB ratio (see Beatty et al., 1999; Cheng and McNamara, 2000). Also, the investment bank Lehman Brothers actively uses some sort of combination CCV model that is mainly based on the relationship of certain operating profit measures and the company's assets (see Löhnert and Böckmann, 2005: 421-427). Nevertheless, it remains to be shown empirically for the German market whether certain multi model approaches can yield superior valuation results as compared to classical single-factor models.

4.3.1.3 Empirical Approaches

A further variant of multi-factor models is the empirical approach. This approach is typically based on regression analysis. In the simplest case, the corporate value (as the dependent variable) is explained by a set of independent fundamental variables. Principally, every set of variables can be applied on the right hand side of the regression equation.[150] Backed by financial theory, however, are only those models that rely on the ideas of the Ohlson model.[151] In most of them price is explained by book value of equity and earnings (and sometimes dividends) (see Herrmann, 2002: 119). In a study conducted for the German market it was found that such a model has high explanation power, especially for the period after 1990 (see Möller and Schmidt, 1998: 495-498).

Additionally, there are other empirical approaches that determine the corporate value in two steps. These approaches aim to explain the value of a certain single-factor model – or its inverse – by a set of variables in the first step. In a second step, this *multiple* has to be applied to the respective basis of reference. In such models, the PE ratio often serves as the dependent variable (see Herrmann, 2002: 114-115). Usually the independent variables are a measure of the systematic risk (the company's beta-factor) and measures of earnings growth.[152] Studies show that this method of determining the value of single-factor models can contribute to improving the valuation accuracy. However, it still allows for mispricing (see Damodaran, 2001a: 294).

The central point of criticism about empirical approaches is that different companies might have different factor sensitivities (see DeFusco et al., 2001: 602). Thus, they are probably priced accurately on average, but there might be noticeable mispricings in single cases. This is a major problem and dramatically reduces the applicability of empirical models in CCV. A second point of criticism is that it is doubtable as to whether there is really a linear relationship between the corporate value (respectively the single-factor model) and the independent fundamentals (see Herrmann, 2002: 120; Damodaran, 2001a: 296). That critique seems to be legitimate.[153] However, single-factor models rely on the same assumption and even worse, they do not only imply a linear relationship, but also imply that the valuation line passes the origin. Therefore, empirical approaches cannot be seen as inferior to classical single-factor models for that reason.

There has also been other criticism about empirical approaches (lack of theoretical foundation, big and heterogeneous samples, etc.) (see Herrmann, 2002: 120-121). However, most of it is not of major relevance.[154] Nevertheless, there is room for improvement. Therefore, the next section tries to establish a multi-factor

150 These models are very similar to the Fama-French multi-factor model and certain variants of the arbitrage pricing theory; for these models see e.g. Fama and French (1993); Copeland et al. (2005: 873-875); as regards the German market see Ziegler et al. (2003).

151 Most studies do not properly account for the "information not yet captured by the accounting system". Therefore, they cannot be seen as *actual* tests of the Ohlson model, see Lo and Lys (2000a).

152 See Herrmann (2002: 115); Beaver and Morse (1978: 72); Zarowin (1990: 448); Damodaran also includes the payout ratio as an independent variable into his analysis, see Damodaran (2001a: 294).

153 See sections 4.2.2.1 and 4.3.2.2 for an in-depth discussion about linearity in CCV.

154 See section 3.2.3.

model that overcomes most of the shortcomings of the single-factor models and the multi-factor models presented thus far.[155]

4.3.2 Derivation of a Two-Factor Model Based on Book Value and Earnings

To theoretically derive a two-factor CCV model one has to rely on existing theoretical works about the influence of accounting variables on the corporate value. As has been outlined above, the only reasonable theoretical basis for a model that includes more than one explaining variable can be found in the RIV model and the studies of Ohlson. There are no other plausible closed-form theories that manage to link a set of accounting variables to the price, respectively the value of a company. However, there are some theoretical ideas that might add to improve the valuation accuracy of such a model under certain circumstances. One of them is the cognition that book value is assumed to be the more value relevant the lower the profitability of the company is. Another one is that, contemporaneously, earnings' value relevance is expected to decrease as profitability declines.

Based on these perceptions, the derivation of the two-factor model proceeds in two steps: First, independent of the operating strength of the target company, a simple CCV model in the sense of Ohlson is developed that explains the corporate value as a function of book value and earnings. Second, this basic model is expanded by considering the relative change in value relevance of earnings and book value in the case of a company's profitability shift. This is done by incorporating management's possibility to abandon and sell firm assets (see Berger et al., 1996) or to adapt firm assets to a different use (see Burgstahler and Dichev, 1997) if the company has operating problems.

With the development of this two-factor model two goals are achieved. (1) The use of this model should explicitly allow for the determination of the value of stocks. Thus, the model should principally be applicable in practice. (2) Based on the model it should be demonstrated how best to combine earnings multiples and book value multiples. Of course, in valuation practice appraisers perform CCV on the basis of both kinds of ratios – those with book value and those with earnings as a basis of reference. The weightings they assign to the results, however, are not necessarily consistent with economic reality. Thus, the development of the two-factor CCV model should also help to reasonably assign weights to the corporate values determined on the basis of the PE ratio and the PB ratio.

4.3.2.1 Recursion Value

The recursion value of a company is defined as the corporate value that should be expected under the assumption that the company continues to apply its current business technology to its resources (strict going concern assumption) (see Burgstahler and Dichev, 1997: 188). The recursion value model is built on the

155 The following outlines are mainly based on Meitner (2003a) and Meitner (2004).

concepts provided by the RIV model and the Ohlson model. However, the model should be adapted to real capital market settings and therefore some of the limiting assumptions of the Ohlson model have to be rearranged. A methodological advantage of the recursion model – as compared to the Ohlson model – is that fewer theoretical requirements are needed because persistence parameters and discount factors can be drawn directly and in an aggregated form from the comparable companies' and capital market's data.[156] Additionally, the framework provided by Ohlson is extended in allowing for transient or even lasting growth of earnings.

Of course, some of the existing empirical and theoretical work has already taken into consideration the idea of the joint value relevance of book value and earnings in CCV. However, the proceeding applied here widely differs from the approaches used in these studies. In contrast to the Cheng and McNamara multi-model approach (see Cheng and McNamara, 2000: 360-362) it is not assumed that PE and PB ratios are equally weighted. Moreover, it is not assumed that companies which earn their cost of capital necessarily have a market value equal to book value (as similarly assumed in an approach provided by Merrill Lynch [see Kames, 2000: 106-108]). Assuming the identity of market value and book value in such a situation would not be appropriate because most accounting systems rather tend towards a conservative accounting. Therefore, for the recursion value model it is assumed that the market value is a function of regression-weighted book values and earnings.[157]

The model develops as follows: Starting from Equation 3.9, the "information not yet captured by the accounting system" is neglected. This step seems to be reasonable because this information can be completely incorporated into the weights of book value and earnings if it is assumed that the comparable companies are subject to the same kind of information (see Penman, 1997: 9-12). However, to account for the possibility that some firm-specific short-term information exists which is incorporated in near-term future earnings, expected next period earnings are used rather than trailing earnings. Consequently, the term $k \cdot \left(\pi \cdot NI_t - DIV_t\right)$ is substituted with $\beta \cdot E\left[NI_{t+1}\right]$ and, thus, it is further assumed that future dividends do not affect the market value of equity because of the clean surplus relation.[158] In doing this substitution the risk of suppressing any information, released

156 In empirical tests of the Ohlson model, persistence parameters are not drawn from comparable companies but from historical data of the target company. Ohlson himself assumes the persistence parameter to be constant over time; see Ohlson (1995): 686, Ohlson (2001: 110). However, from an appraiser's point of view this approach seems to be highly debateable, because of quick changing market conditions and industry characteristics.

157 The concept of the model applied here is comparable with the concept of one of the models in Beatty et al. (1999); a similar model has been presented by Ramakrishnan and Thomas (1992: 442-447).

158 It can be shown that future dividends are irrelevant even if the accounting is not perfectly clean surplus and even if the "value – reference variable" association is non-

after the disclosure of current earnings, can be avoided. This is especially important because the date of valuation rarely equals the date of financial statement disclosure.

In contrast to the Ohlson model the weightings correspond to earnings but not to discounted earnings. As a consequence, the earnings' weighting factor β does not necessarily need to be in the range between 0 and 1. As mentioned already, the book value's weighting factor may also differ from values between 0 and 1 because of conservative accounting and the possibility of short-term earnings growth attributed to the size of the firm's assets. Consequently, the weights of book value and earnings are not constrained to sum up to 1.

It should be noticed that from an economic point of view even negative weights for the book value of equity might be reasonable. In this case, certain underperforming companies have a negative recursion value even if they have positive expected earnings; ceteris paribus their absolute recursion value is indirectly related to book value. This phenomenon can occur e.g. in an industry that tends to a natural monopoly and where earnings can be viewed as a proxy for market share. In contrast, negative weights for expected earnings do not make any economic sense. Instead, this would be a sign for bad comparable company selection.

One of the advantages of the regression-weighted approach is that it allows for a better description of the influence the two reference variables book value and earnings have on the equity value. However, there is also a methodological advantage when using the regression-weighted approach: The recursion value model can easily be attuned to the option to liquidate or to adapt by changing the strategy in the case of bad current expectations of business activities discussed below.[159]

In the final recursion value model the market value of equity of the company at time t, assuming that it continues its current business activities ($V_{rec,t}$) can be expressed as follows:

$$V_{rec,t} = \alpha \cdot EQUT_t + \beta \cdot E\left[NI_{t+1}\right] \tag{4.5}$$

where α, β are the respective regression weights for the current book value and the expected earnings.

Since the term $\left(\pi \cdot NI_t - DIV_t\right)$ from Equation 3.9 approximately equals $k_e^{-1} \cdot E\left[NI_{t+1}\right]$ in the settings of the recursion value Equation 4.5 presented above[160], the following statements can be made regarding the development of fu-

linear due to the existence of abandonment options, see Yee (2005). This finding is of paramount importance for the following sections.

159 See next section.

160 While the Ohlson model assumes risk neutrality of all investors, see Ohlson (1995: 665-666), this model also works in the more realistic settings of investors' risk aversion. Strictly speaking, a model extension that incorporates risk into the discount rates "lacks theoretical appeal", Ohlson (1995: 680), but this is of minor importance here because the weightings are empirically determined and not theoretically derived.

ture earnings: If $\hat{\beta} \cdot k_e > 1$, then future earnings are expected to grow perpetually. If $\hat{\beta} \cdot k_e = 1$, then earnings remain constant over time; if $\hat{\beta} \cdot k_e < 1$, then future earnings approach a certain target-return on equity (ROE) over time. This ROE asymptotically equals $(\beta \cdot k_e + \alpha) \cdot k_e$

since $\left(V_{rec,t} \cdot EQUT_t^{-1}\right)_{E[ROE_t]=k_e} = \beta \cdot k_e + \alpha$.

In this context it should be noticed that the term $(\beta \cdot k_e + \alpha) \cdot EQUT_t$ does not necessarily equal shareholder funds because sometimes earnings approach a certain $ROE > k_e$ in the long-run, depending on the degree of competition in the respective industry. In such industries (e.g. price searcher markets with high entry barriers such as the Life Science industry or the Machinery industry) companies can infinitely earn abnormal profits (see Gwartney and Stroup, 1997: 554-561).

Some academic research shows that market value is not a linear function of book value and earnings (see e.g. Penman, 1996; Penman, 1997). However, it is assumed in the first step that *recursion value* is indeed a linear function of these two variables. This assumption seems reasonable as long as peer group companies are affected in the same way by the same drivers of the "earnings/book value" relevance cube[161], independently of their actual ROE. This seems to be a reasonable assumption for companies that are characterised by the same industry classification and similar financial leverage but, in fact, it still remains to be shown whether this model allows comparable companies to be selected based on these two criteria. However, even if these two criteria suffice in peer group selection, one should consider that well performing companies probably have more power within the industry than poor performing companies. Regarding the earnings–book value relevance cube, this could lead to a higher value relevance of earnings along with a lower value relevance of book value for these companies as compared to low-ROE companies.

It is also important to note that – since the model sometimes assigns positive recursion values to companies with current negative earnings – the linear relationship only holds if the market discounts negative earnings with the same rate as it discounts positive earnings. This use of a unique discount rate has sometimes been challenged (see Berry and Dyson, 1980; Booth, 1982) but more recently also supported in theoretical studies (see Ariel, 1998). Nonetheless, as will be shown below, the problem of potentially differing discount rates is of minor importance for the final CCV two-factor model because the recursion value will be modified to account for the option to reorganise the company.

---

161 See section 3.2.1.3.

4.3.2.2 Option to Reorganise the Company

As has been outlined in section 3.2.1.3, the amount of assets available does not only play a major role in determining future earnings assuming the company continues its current business activities but also in determining the company's value in case of abandonment or change of current business activities. It is important to note that for the present analysis abandonment can be understood as total abandonment of the whole firm but also as a partial asset sell-off, like spin-offs or equity carve outs. Likewise, adaptation does not necessary mean full and immediate change of current activities. It can also mean to change the current form of business activities bit-by-bit or even to change only a part of current activities.[162]

For the very most part of companies, a potential liquidation or adaptation takes place at a future point in time. As a matter of consequence, the exit values for these scenarios are generally unobservable at the time of valuation. External appraisers, on the whole, have no direct access to data with which they can determine the real values. Instead, they must rely on publicly available accounting figures that are close to these values. Even if neither liquidation value nor the value of firm resources adapted to some superior use necessarily conforms to book value (see Sieben and Maltry, 2005: 399; Richter, 2002: 307-311), the differences should be assumed to be typically small. This is obvious for the liquidation value because both values imply that the company is not viable as a going concern (see Berger et al., 1996; Barth et al., 1998). However, it also holds approximately for the adaptation value: Adaptation value is an unknown figure not only for external appraisers but also for company insiders because though management knows that it has to adapt firm resources there is some initial uncertainty concerning the specific use these assets should be adapted to. Balancing the benefits of a possible superior use of assets with the cost of adaptation, book value should be a prudent first estimate of adaptation value.

It should be noticed that even if it seems reasonable to proxy, in the first step, the amount of assets available to sell or to adapt by book value, this might be inappropriate dependent on the level of information: If in a specific valuation setting the appraiser knows the real reorganisation value – which by definition includes both factual adaptation and sale of assets – or a better proxy, this new value should be substituted for book value. However, for reasons of simplicity and to keep the model as general as possible, book value serves as a proxy for both exit values.

To develop the reorganisation value function, book value of equity (and not book value of total assets) is used because it is the accounting figure relevant for shareholders. This is consistent with the proceedings of the empirical study by Barth et al. (1998) and allows better accounting for the senior position of debt in

---

[162] A well-known example for this is the metamorphosis of the German Mannesmann AG in the late 1990s (before the merger with Vodafone). The company changed its business focus from pipe producing to coal, iron and steel manufacturing to finally become a pure play telecommunication company. Another example for gradual adaptation is the conversion of the German industry group Preussag AG into the tourist company TUI.

case of bankruptcy. To make this reorganisation value comparable to the recursion value, expected next period book value (assuming the company pays no dividends) is used. Thus, the reorganisation value of the company at time t ($V_{reo,t}$) can be expressed as follows:

$$V_{reo,t} = EQUT_t + E\left[NI_{t+1}\right] \tag{4.6}$$

The assumption that reorganisation value depends on the next period's expected earnings is not only important to reach consistency with recursion value, it is also economicly plausible since the process of adaptation does not start immediately but (if at all) at any time in the future. Even if the process starts soon after the date of valuation, it will usually last several months, so that next period's earnings affect stockholders equity and therefore directly change the amount of available resources.

This function of reorganisation value has a null value for $E\left[NI_{t+1}\right] = -EQUT_t$, indicating that the company has no assets left when it loses all its equity through next year's negative earnings.

The relevance of the reorganisation value in business valuation is especially high if the company is very likely to adapt firm resources. The probability of adaptation, in turn, is dependent on the current operating performance and efficiency of the company: If a company's business is flourishing and current earnings are high relative to book value, it is very unlikely that the management will give up its current operations. In this case reorganisation value is obviously of little significance compared to recursion value. Contrary, if a company's current activities are not satisfyingly successful the probability of abandonment or strategy change is higher. In the latter case the reorganisation value plays a major role in determining the company's market value.

The threshold of change to adaptation is reached when a company can get more out of an asset or a certain group of assets by selling it or adapting it than by continuously applying it to its current use, i.e. when NPVs in the case of adaptation exceeds NPVs in case of continued use. In the present case of multi-factor CCV management reorganises firm resources if $V_{reo}>V_{rec}$.

However, since valuation circumstances are typically characterised by uncertainty and information asymmetry and, thus, (especially) the reorganisation value cannot be exactly determined, the real position of the threshold of reorganisation is ex ante indeterminable. Whether firm assets will be liquidated in the future depends on the probability of default (externally induced liquidation) or on management's discretion (internally induced liquidation). Similar uncertainties hold for adaptation: Whether firm assets are adapted to a different use depends on management discretion based on management's estimates about the future of current business activities and about the success of adaptation.

In fact, there is always a certain probability of business reorganisation – this probability is high if current earnings are low and it is low if current earnings are high. Thus, the fair value of a company is a function of both recursion value and reorganisation value because management has an *ongoing option* to either continue its present operating activities or to adapt its resources to different uses.[163] Consequently, not only recursion value but also the value of the option to reorganise current business activities (in this case the option is an American style long put) should be reflected in the market value of the company. If this option is not accounted for, it would mostly understate the value of companies with very low or negative earnings.

4.3.2.3 Valuation Model

To derive a model for the target company's market value of equity – with regard to the existence of recursion value as well as the option to switch to reorganisation value – Equations 4.5 and 4.6 first have to be restated:

$$V_{rec,t} = \alpha \cdot EQUT_t + \beta \cdot E\left[NI_{t+1}\right]$$

$$V_{rec,t} \cdot EQUT_t^{-1} = \alpha + \beta \cdot ROE_{t+1} \quad (4.7)$$

and

$$V_{reo,t} = EQUT_t + E\left[NI_{t+1}\right]$$

$$V_{reo,t} \cdot EQUT_t^{-1} = 1 + ROE_{t+1} + \varepsilon \quad (4.8)$$

where ROE_{t+1} is short for $E\left[NI_{t+1}\right] / EQUT_t$, denoting the expected return on equity of period 1, and ε is the normally distributed additive error with expectation value $E[\varepsilon]=0$ and standard deviation $\sigma[\varepsilon]=\sigma_{reo}$; i.e. $\varepsilon \sim N(0, \sigma_{reo})$. The error term is necessary since the real reorganisation value is unknown.[164]

The restatement of both equations facilitates a graphical visualisation of the relationship between earnings, book value and market value in a two-dimensional space. See Figure 29 where $V \cdot EQUT_t^{-1}$ (i.e. the PB ratio) is plotted on the ordinate and ROE_{t+1} on the abscissa.

To determine the recursion value weights α and β of Equation 4.5, the PB ratio is regressed linearly on ROE. This proceeding allows for the illustration of the relative value relevance of the two accounting attributes (book value and earnings)

---

163 As regards the option-style character of equity especially for distressed companies, see Damodaran (2002: 817-830). As regards the optimal date of reorganisation under uncertainty, see Richter (2002: 300-306).

164 There is no error term in the recursion value formula because it is assumed that earnings and book value are satisfyingly value relevant if applied jointly.

and, implicitly, the expected earnings future development and persistence for an average peer group company.

Unfortunately, the regression based aggregation method suffers from a major problem: In general, stock prices are influenced by interdependencies between recursion and (potential) reorganisation. Thus, to identify pure recursion value, a set of companies is needed whose stock prices are largely unaffected by the abandonment or adaptation option. Consequently, for the regression only data of companies with an expected ROE exceeding the cost of equity or at least close to its cost of equity are used.[165] That implies that the earnings/stock price relationship is expected to be linear only if a company earns at least a certain return on equity. Of course, a sufficiently large set of profitable companies is needed to run this regression accurately.

A steep slope of the regression line indicates a high persistence and value relevance of current earnings. In this case, the regression line resembles the PE ratio line. If, in contrast, the regression line is very flat and its course is almost parallel to the ROE axis, then current abnormal earnings are a bad indicator of future cash flows because future earnings are expected to converge to the industry-wide ROE. In this case, current book value of equity significantly influences the market value of equity. As a consequence, the regression line resembles the PB ratio-line.

Figure 29 exemplifies the recursion value and the reorganisation value for an average company in the German telecommunication industry (date: 14-11-2003). Expected ROEs are calculated with I/B/E/S earnings forecasts 2003 drawn from Bloomberg. Therefore only companies with an I/B/E/S forecast available are included. Each data point in the figure represents a peer group company. The diamonds (♦) denote companies with an expected ROE less than 9 % and the stars (*) stand for companies with an expected ROE greater than 9 %.[166] The dashed recursion value line is the result of a regression run with the star-shaped data points while the thin continuous line represents reorganisation value.

To finally establish the functional relationship between the PB ratio and the future ROE, the expected maximum of the reorganisation value and the recursion value is computed.[167] In doing so, recursion is considered and also the ongoing option to switch from recursion to reorganisation can be incorporated into the model:

$$PB = E\left[\max\left(V_{reo}, V_{rec}\right)\right] \tag{4.9}$$

---

165 The respective threshold effectively depends upon the slope of the recursion value line, the moments of ε, and the point of intersection between $E[V_{reo}]$ and V_{rec}. That is why in certain "low reorganisation value" industries even for firms with negative earnings the abandonment option is sometimes irrelevant. This was the case e.g. in the IT-sector in the late 1990s.

166 As we will see later – for the selected σ_{reo} – all companies with a *ROE* bigger than 9 % are nearly unaffected by the reorganisation option.

167 This proceeding is based on the idea of Trigeorgis (1996: 12), and Burgstahler and Dichev (1997).

Fig. 29. Recursion value and reorganisation value

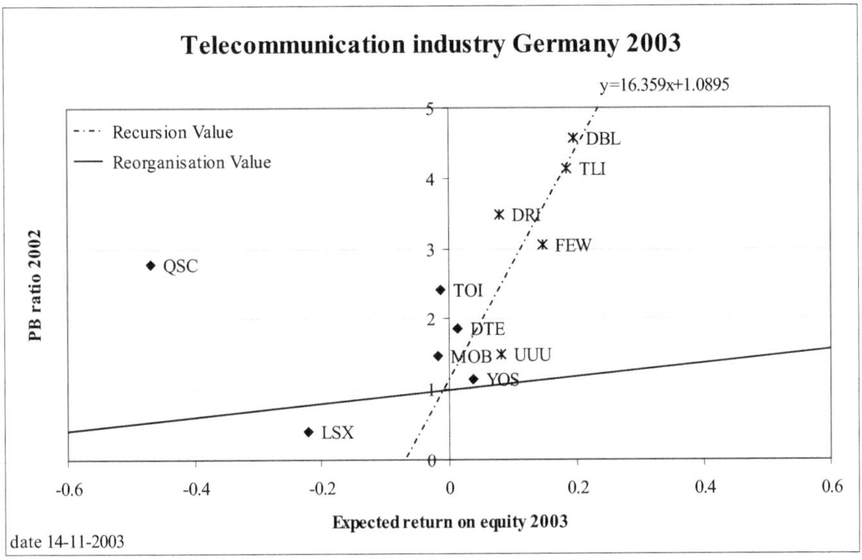

Source: ZEW, Bloomberg.

$$PB = \int_{-\infty}^{+\infty} \max\left\{1 + ROE + \varepsilon, \hat{\alpha} + \hat{\beta} \cdot ROE\right\} f_{\varepsilon}(\varepsilon) d\varepsilon \tag{4.10}$$

where $f_{\varepsilon}(\varepsilon)$ denotes the probability density function of the normally distributed additive error ε. Considering that V_{reo} is high relative to V_{rec} for low-profitability firms yields:

$$PB = \int_{-\infty}^{\infty} \left[\left(\hat{\alpha} + \hat{\beta} \cdot ROE\right) \cdot 1_{\left[-\infty, d\right]}(\varepsilon) + \left(1 + ROE + \varepsilon\right) \cdot 1_{\left[d, +\infty\right]}(\varepsilon) \right] \cdot f_{\varepsilon}(\varepsilon) d\varepsilon \tag{4.11}$$

where $$1_{[a,b]}(\varepsilon) := \begin{cases} 1, \text{ if } \varepsilon \in [a,b] \\ 0, \text{ if } \varepsilon \notin [a,b] \end{cases}$$

and the parameter *d* is defined as the realisation of ε which leads to investors' indifference between recursion value and reorganisation value. It is obtained by setting V_{reo} equal to V_{rec} and solving for ε which yields:

$$d \equiv V_{rec} - E\left[V_{reo}\right] = \hat{\alpha} - 1 + \left(\hat{\beta} - 1\right) \cdot ROE_{t+1} \cdot$$

Rearranging the main equation leads to:

$$PB = (1+ROE) + \left(\hat{\alpha} - 1 + (\hat{\beta} - 1) \cdot ROE\right) \int_{-\infty}^{d} f_{\varepsilon}(\varepsilon)\, d\varepsilon + \int_{d}^{+\infty} \varepsilon \cdot f_{\varepsilon}(\varepsilon)\, d\varepsilon \qquad (4.12)$$

Since $\int_{-\infty}^{d} f_x(x)\,dx = \Phi\left(\frac{d}{\sigma}\right)$ and $\int_{d}^{+\infty} \varepsilon \cdot f_{\varepsilon}(\varepsilon)\,d\varepsilon = \sigma \cdot \varphi\left(\frac{d}{\sigma}\right)$ with $\Phi(x)$ denoting the cumulative standard normal distribution of *x*, and $\varphi(x)$ denoting the standard normal distribution of *x*, the final function is:

$$PB = E\left[V_{reo}\right] + d \cdot \Phi\left(\frac{d}{\sigma_{reo}}\right) + \sigma_{reo} \cdot \varphi\left(\frac{d}{\sigma_{reo}}\right) \qquad (4.13)$$

with $E\left[V_{reo}\right] = 1 + ROE_{t+1}$

The resulting model is practically applicable, i.e. all the necessary parameters can be drawn from previous calculations or directly from the capital market.[168]

In this model high importance should be attached to the standard deviation of the reorganisation value error term σ_{reo}. Since it is ex ante not clear to what specific use the firm assets should be adapted, it seems to be reasonable to proxy this standard deviation by the volatility of a broad market index (e.g. the German DAX). To determine the period up to expiry of the reorganisation option is obviously more difficult. Management certainly always has the possibility to reorganise, but it is not economicly sound to use an infinite time horizon for valuation due to the timely limited forecast periods of market participants. More so, the author may think that an appropriate assumption would be that the market prices this option with a time to expiration equal to the period for which appraisers usually perform detailed future cash flow forecasts in common DCF valuation models; i.e. up to 5 years (see e.g. Copeland et al., 2000: 234; IDW, 2002: 61). However – just as in the case of the determination of the reorganisation value – if appraisers can predict the parameters of this option better they might use a deviant time to expiration here, too.

Figure 30 illustrates the two-factor CCV model. The thick curve represents the target company's PB ratio dependent on expected next year *ROE*. The parameters α (1.0895) and β (16.359) equal the weights of the regressional determination of the recursion value, σ_{reo} denotes the (unannualized) historical 5 year volatility of the DAX.[169] It can be seen that the data points of underperforming companies like the Deutsche Telekom AG (DTE), the Mobilcom AG (MOB) and the LS Telcom

168 A detailed calculation of this two-factor model can be found in Appendix 7.1.

169 The historical volatility of the DAX was calculated based on daily prices for the last 5 years.

AG (LSX) are not far away from the thick curve. However, the results might lead to the supposition that the reorganisation value of the LS Telcom AG is overstated in our model, i.e. that it is smaller than book value in reality. Even worse, the model obviously fails to accurately value the QSC AG (QSC). Nevertheless, the problems associated with these inaccuracies can most likely be mitigated by a more detailed analysis of the respective companies (i.e. more thorough determination of the reorganisation value of LS Telcom and clear identification of the reasons for the relatively high PB ratio of QSC).

Fig. 30. The two-factor comparable company valuation model

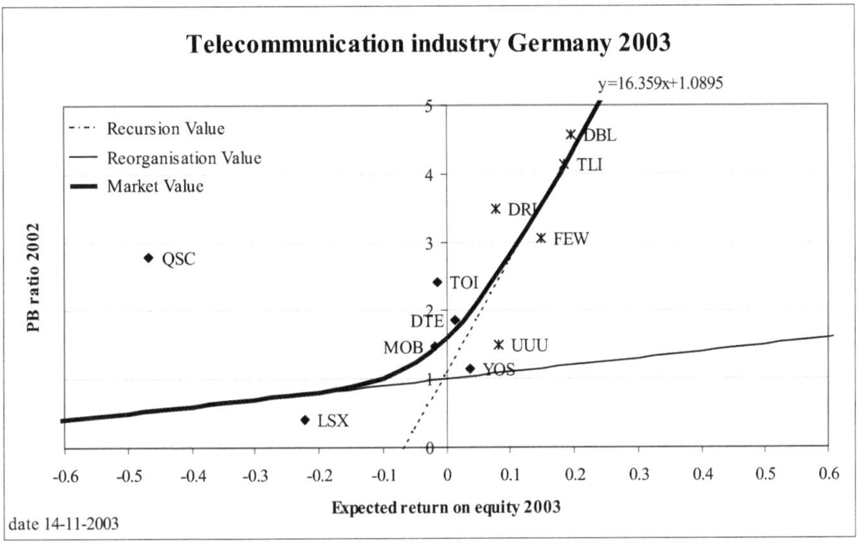

Source: ZEW, Bloomberg.

4.3.2.4 Selection of Comparable Companies

Finally, this sub-section deals with the question as to whether the selection of comparable companies based on industry selection – as it has been done in the previous section – is really appropriate. This issue will be addressed below, considering the fact that the two-factor model only requires selecting comparable companies for the determination of the recursion value.

First, it is necessary to equate the recursion value function (Equation 4.5) to the Ohlson model (Equation 3.8) assuming that recursion value and price are identical. The left hand side of the resulting equation denotes the empirically determined value while the right hand side is the theoretical valuation model:

$$\alpha \cdot EQUT_t + \beta \cdot E\left[NI_{t+1}\right] = (1-k) \cdot EQUT_t + k \cdot (\pi \cdot NI_t - DIV)_t + \lambda \cdot \nu_t \qquad (4.14)$$

In the next step, the right hand side of the equation has to be adjusted in three respects. Consistent with the proceeding in section 4.3.2.1, one adjustment is the neglection of the "information not yet captured by the accounting system" ν, and the second adjustment is the substitution of current earnings and dividends by expected future earnings. Additionally, since accounting book value is only a noisy estimate of economic book value because of accounting restrictions (see Kuhlmann, 2005: 101), a correction factor w is included. This correction factor is a measure of deviation between both kinds of book value and strongly depends on the accounting system used. All of these adjustments change the former equation to:

$$\alpha \cdot EQUT_t + \beta \cdot E\left[NI_{t+1}\right] = (1-k) \cdot w \cdot EQUT_t + k \cdot \eta \cdot E\left[NI_{t+1}\right] \qquad (4.15)$$

In this equation, k is similar to the parameter k in Equation 3.8 and denotes a measure of persistence of earnings. The variable η is a discount factor that is similar – but not identical – to the variable π in Equation 3.8. The difference between the two variables is that η is not based on risk free discount rates but rather represents a factor that accounts for the operating and financial riskiness of companies' cost of capital. It is assumed that companies with identical cost of capital should also exhibit similarity in the variable η.

The main conclusion to be drawn from Equation 4.15 is that companies are expected to have the same (empirically determined) α and β if they are identical in the three parameters k, w and η. The variable w is identical for all companies that use identical accounting procedures – which could be largely assumed for companies that apply the same accounting standards – and η is at least similar for companies in the same industry with the same capital structure since these companies usually have a similar cost of capital[170]. Additionally, following the design of the Ohlson model, the variable k is identical for all companies that exhibit the same persistence parameter with regard to abnormal earnings. The persistence parameter, in turn, is a function of whether abnormal earnings will revert to the normal level over time and – if they do – of the speed of this reversion. As has been extensively discussed in section 3.2.1.3, the persistence of abnormal earnings is crucially determined by the industry to which the company belongs. Consequently, the level of persistence can be assumed to be largely the same for all companies in one industry.

Putting all this together emphasizes that the selection of comparable companies can be based on the two criteria "same industry classification" and "similar capital structure". The major perception here is that growth perspectives are not explicitly a criterion of peer group selection (as opposed to single-factor models). Obviously, the joint inclusion of earnings and book value (and therefore the implicit in-

---

170 The uniformity of the cost of equity within one industry has been highlighted by the Deutsche Boerse AG (German Stock Exchange) until 2003 by publishing industry betas together with the CDAX industry sub-indexes on its homepage. This service has been discontinued in the course of the new-segmentation of the equity market and the associated disappearance of these sub-indexes. However, other providers still calculate and publish industry betas, e.g. DIT (2004: 6-7). A critical assessment of this beta uniformity within industries can be found in Timmreck (2004: 65).

clusion of ROE in the valuation model) reduces the similarity requirements from growth down to persistence. This is not astonishing, given that ROE (i.e. profitability) is one of the main drivers of the financial benefits' growth rate.[171] While at first glance this is only a minor change in the selection criteria, the consequences are dramatic: Valuation theory now allows to principally put together a bigger set of comparable companies (the selection is only restricted by "same industry classification" and "similar capital structure") as compared to classical single-factor models (where – at least – initial ROE is an additional similarity requirement).

4.3.2.5 Conclusions Regarding the Two-Factor Model

From a *theoretical point of view* the derived two-factor model especially provides four improvements as compared to common single-factor models, in particular:

- The two-factor model allows for the depiction of expected future earnings development more economicly sound than single-factor models do. This is the case because it must be assumed that book value and earnings have higher value relevance than each of these reference variables alone. To put it more precisely, in most cases the two-factor model explains more of the corporate value than a single-factor model does. Therefore, from a theoretical point of view the valuation accuracy of the two-factor model should be generally higher.
- By including the option to reorganise the company, the two-factor model can conclusively assign positive stock prices to currently negatively performing companies.
- Following valuation theory, the two-factor model allows for a principally high number of comparable companies. This is of major concern when valuing companies in low populated industries. It is, however, also important in high populated industries since the factual degree of market efficiency and minor deviations in the similarity of the operating business of companies requires the number of reasonable comparable companies to be as big as possible in order to even out these differences.
- For appraisers that follow the practice-oriented approach to select comparable companies (this approach centres around the "industry classification" as the main comparability criterion and enjoys widespread popularity in valuation practice), the two-factor model provides more ex ante valuation accuracy than the single-factor models. This is the case because the practical approach is rather consistent with the theoretical selection requirements of the two-factor model than with those of classical single-factor models.

However, what appears to be a major problem in practical applications of the derived two-factor model is the determination of the time to expiration of the option to reorganise the company (which is set at 5 years for the example above). Additionally, to proxy the reorganisation value by book value of equity might be a rather conservative move in many cases. This all suggests that there is still room

---

[171] See section 4.2.2.5.

for improvement of this model. However, it remains to be tested empirically[172] whether the derived two-factor model can compete with or even outperform other CCV models.

172 See section 5.3.

5 Empirical Study

The empirical examination is split into two parts. The first one is about *value relevance*. Value relevance is a measure of how certain reference variables can explain stock prices. It is important to note that value relevance only focuses on the relationship between the target company's accounting variables and its stock price. Therefore, value relevance studies ignore that CCV requires a set of comparable companies to project a stock price onto the target company. To put it more precisely, a study about value relevance will shed light on how best to perform CCV in situations where there is no lack of comparable companies. As has been outlined already, the selection of comparable companies is an important but difficult and time consuming task in valuation practice. However, in a broad empirical study – such as the study at hand – it is not possible to accurately and individually select comparable companies for each valuation case. The value relevance study accounts for that and allows drawing important conclusions about CCV without an in-depth analysis about every potential comparable company. Thus, this approach is very important because it may help appraisers to focus on economic influences when choosing the valuation model and renders advices on how to behave in real valuation settings.

The second part of the empirical study is an examination of the *pricing accuracy* of different models assuming that appraisers apply the practice-oriented approach to collecting the set of comparable companies, i.e. assuming that appraisers strongly focus on industry classification. This analysis is performed because the practice-oriented approach to selecting comparable companies still enjoys widespread popularity amongst investment professionals. Additionally, many empirical studies about CCV have applied this approach in the past. This study about pricing accuracy aims at particularly showing how the two-factor model that was derived in section 4.3.2 has performed in the past.

5.1 Data

5.1.1 Sample Selection

The empirical analysis is based on a panel of stock exchange listed German companies, covering the years 1998-2003. The sample only contains German companies because foreign companies are exposed to a differing regulatory and tax environment and, thus, cannot be regarded as comparable. The basis for the sample selection is the stock index CDAX. This index encompasses all German companies from the Deutsche Börse AG market segments Prime Standard and General Standard. The index represents the entire range of the German equity market. Only companies that were continuously listed over the sample period (resp. listed from IPO until the end of the sample period) were included in the analysis.

Table 2. Sample composition

Firm year observations 1998-2003 (i.e. German companies listed in the CDAX)[a]	about 2,970
– banks and insurance companies	
– observations of companies that disclose consolidated financial statements according to German GAAP (and with no reconciliation to IAS/IFRS or US-GAAP available)	
– companies with a balance sheet date other than December 31	
– observations of companies that filed for insolvency	
– observations of companies for which data were not readily available	
– observations of companies with stock price >100 (=7.8%)	
= **Basic sample**	**928**
Basic sample if information about industry structure is required	810
Basic sample if information about creditworthiness is required	675

[a] Source: Deutsche Börse Factbooks 1998-2003; no exact number of listed German companies was available for the year 1999.

Consistent with prior research, banks and insurance companies were excluded from the sample. To ensure comparability of data, observations of companies that do not prepare consolidated financial statements according to IAS/IFRS or US-GAAP are eliminated as well as observations of companies that have a balance sheet date other than December 31. Observations of companies which filed for insolvency, but which were not delisted at the balance sheet date, were also excluded because of the risk of irrational pricing. Additionally, several observations had to be deleted because stock price and/or financial data were not readily available for them. Finally, to ensure that estimation results are not sensitive to extreme values, observations with stock prices above 100 Euro were removed from the sample.

After all these exclusions, a basic sample of 928 observations remained for the analyses. However, this basic sample could not be applied in all examinations. For some tests, further information was necessary which again reduced the number of observations. For example, analyses that are based on information about industry structures could only be performed with 810 observations; if data about creditworthiness were required, only 675 observations were available. Table 2 summarises the sample selection process.

5.1.2 Variables

Most input variables stem from financial statement data sources and stock price databases. Additionally, information about industry structures and creditworthiness of companies as well as macroeconomic information was required for this empirical analysis.

The source of *stock price data* is the Karlsruhe capital market database (Karlsruher Kapitalmarktdatenbank, KKMDB), which is a part of the German Finance Database (Deutsche Finanzdatenbank). The KKMDB contains stock price information on all stocks traded on the Frankfurt Stock Exchange since 1960, and of several stocks traded on other German stock exchanges as well. The KKMDB also calculates and publishes a reference index, which is called the German Stock Price Research Index (Deutscher Aktien-Forschungsindex, DAFOX). For this examination, data from 1997-2004 were used. Stock prices and index prices are observed once a day. All these data are corrected for dividend payments, equity issues, stock splits and other factors that cause stock price changes due to technical reasons.

The primary source of *financial statement data* is Hoppenstedt's Balance Sheet Database. This database contains complete information about balance sheets, income statements and cash flow statements from many large and medium-sized German corporations that do not operate primarily in the financial services industry. Hence, banks and insurance companies are not included. An additional source of financial statement data are the annual reports of companies. For the present examination, only information relating to consolidated balance sheet, income statement, and cash flow statement during the financial years 1998-2003 were used.

The analysis of *creditworthiness* of companies is based on credit ratings provided by Creditreform, the leading rating agency in Germany. The Creditreform database contains information about creditworthiness and solvency of many Western German firms from all business sectors.

For the analysis of *industry structure*, data from a survey on the innovation behaviour in Germany called the Mannheim Innovation Panel (MIP) and data from the ZEW Start-Up Panel were used. The MIP is conducted every year by the ZEW on behalf of the German Ministry for Education and Research. The methodology and questionnaire of the survey, which is targeted at enterprises with at least five employees, is comparable to the Community Innovation Survey (CIS) conducted every four years by Eurostat. The ZEW-Start-Up panel contains firm level information on about 2,090,000 firms in the West and 795,000 firms in the East of Germany for analyzing business start-ups, growth, and insolvencies.

Furthermore, macroeconomic input-output-tables published by DESTATIS and the Expert Opinion of the Monopoly Commission (Hauptgutachten der Monopolkommission) 2001 served as sources of information. Other macroeconomic variables are drawn from DESTATIS or provided by the ZEW.

5.2 Value Relevance[173]

5.2.1 Previous Empirical Results

The finance literature offers a vast range of value relevance research. Many academics have devoted considerable effort on this topic since Beaver (1968) and Ball and Brown (1968) have set the starting point for this stream of financial research. Therefore, the presentation of a complete literature survey would go by far beyond the scope of this work. However, a broad overview of the value relevance literature can be found in Holthausen and Watts (2001) and Kothari (2001). Möller and Hüfner (2002), in turn, provide an excellent insight into value relevance issues with special regard to the German capital market.

Below, a tabular overview of the most important results concerning research questions that are addressed in the present study is shown. Additionally, two recently conducted studies – which set the methodological standard for the present examination – are discussed in more detail: the examinations of Brief and Zarowin (1999) and of Whelan and McNamara (2004).

Table 3. Overview of the relevant value relevance studies

Value relevance of earnings in general

There is almost no doubt that financial statement variables carry information that is important to value stocks or companies. Especially, a strong relation has been found between stock returns and earnings (e.g. Collins and Kothari, 1989). However, the average measured earnings response coefficient – i.e. the magnitude of that relation – is smaller than predicted by valuation theory. This is explained by the price-lead-earnings phenomenon (Beaver et al., 1980), noise in earnings (Ramakrishnan and Thomas, 1998), the accounting recognition lag (Easton et al., 1992) and the lack of persistence in earnings (Ou and Penman, 1989). Harris et al. (1994) showed that the value relevance of accounting figures is higher under US-GAAP than under German GAAP. Lev and Zarowin (1999) found a declining value relevance of earnings since the middle of the 1970s.

173 All estimations in this study about value relevance are performed using the statistical software STATA release 8.2.

Value relevance of earnings and book value of equity

Most of these examinations are principally based on the seminal works of Ohlson (1995) and Feltham and Ohlson (1995). Many studies find that earnings and book value have complementary value relevance (Kothari and Zimmermann, 1995; Collins et al. 1997; Francis and Schipper, 1999). Book value is typically seen as relatively more relevant for distressed firms than for financially healthy firms (Burgstahler and Dichev, 1997; Barth et al., 1998; Collins et al., 1999). Whelan and McNamara (2004) show that book value is relatively more relevant if firms engage in earnings management than if they do not. Collins et al. (1997) found that the relative value relevance of earnings (book value) has decreased (increased) over the past 40 years. Hung and Subramanyam (2004) showed that book value (net income) plays a relatively lesser (greater) valuation role under German GAAP than under IAS/IFRS.

Value relevance of earnings and cash flows

Mixed results from examinations in large samples: While some studies found that there is virtually no difference between earnings' and cash flows' ability to explain stock returns (Wilson, 1987; Cheng et al., 1996; Cheng and Liu, 1997), others clearly show the superiority of earnings – i.e. of accrual based performance indicators – over cash flows (Dechov, 1994; Sloan, 1996; Charitou et al., 2001; Bartov et al., 2001). All of these results support the theories of rationality and responsibility in management's accounting behaviour and of high quality of earnings in general (see section 3.2.1.4). However, for certain industries (Biddle et al., 1995), if earnings are transitory (Charitou et al., 2000) and if a company is in the start-up or growth stage (Black, 1998) cash flows' explanatory power may sometimes exceed that of earnings.

Convexity in the price-earnings-book value relationship

Burgstahler and Dichev (1997) showed that equity value is a convex function of both earnings and book value. Their reasoning is that book value is a proxy for the value of firm resources, and if a company has low relative earnings it is more likely to exercise the option to adapt these resources to a superior alternative use.

Value relevance in the context of CCV

There are only a few studies that deal with value relevance issues in the context of CCV. Some of the rare examples are provided by Baker and Ruback (1999), Beatty et al. (1999) and Liu et al. (2002).

The study of Brief and Zarowin (1999)

In their paper "The Value Relevance of Dividends, Book Value and Earnings", Brief and Zarowin compare the value relevance of book value and dividends with that of book value and earnings. This examination is representative for a bulk of studies that measure value relevance by the coefficient of determination R^2. In this study two variants of R^2 are applied. The first one is the common R^2 for the simple regressions. The second one is the incremental R^2 as the difference between the R^2 of a multiple regression and the R^2 of a regression that includes all independent variables other than the independent variable under examination. Thus, the incremental R^2 of a specific independent variable has to be understood as the contribution of the specific independent variable to the R^2 of the starting multiple regression. The inclusion of the incremental R^2 allows for a better assessment of the value relevance of accounting figures since it shows how value relevance for certain variables changes if they are applied jointly with other variables (contrary to the pure assessment of value relevance as a stand-alone variable). Brief and Zarowin do not interpret the coefficient values, since these values only have a meaning in the context of value relevance if they are compared over time or between two sub-samples for the same variable.

The study of Whelan and McNamara (2004)

In their paper "The Impact of Earnings Management on the Value-Relevance of Financial Statement Information", Whelan and McNamara examine the *relative* value relevance of earnings and book value in the presence of different sources of earnings management. Contrary to the Brief and Zarowin study, the focus here is on how value relevance changes if certain external criteria change. Therefore, the measure of value relevance is no longer only R^2 but also the change in the coefficients of a multiple regression with book value and earnings as independent variables. The reasoning is that if the slope coefficient of earnings increases relative to the slope coefficient of book value, then the value relevance of earnings has increased relative to the value relevance of book value and vice versa. Whelan and McNamara do not run regressions for different sub-samples but rather include dummy variables for earnings management into the regression for the full sample. The dummy variables have a value of one if companies engage in earnings management and zero otherwise. The advantage of his approach is that the change in value relevance between the sub-samples can be directly calculated from the whole sample regression.

5.2.2 Variable Definition

The following is a list of the variables that are used in the empirical analysis:

Accounting variables[174]

All accounting variables are calculated on a *per share* basis. Some of the accounting variables follow slightly different calculation schemes under IAS/IFRS and under US-GAAP. However, the differences between the respective accounting

174 The financial statement data are identical to the reported data from the consolidated annual reports or those provided by Hoppenstedt's Balance Sheet Database, i.e. no further adjustments for analysis purposes are made.

figures under US-GAAP and IAS/IFRS are typically very small so that it seems to be justifiable to describe each pair by one single variable.[175] The accounting variables are as follows:

NI_{it}	Net income of company i in year t
$EBIT_{it}$	Earnings before interest and taxes (operating income) of company i in year t
$EBITDA_{it}$	Earnings before interest, taxes, depreciation and amortization of company i in year t
$SALES_{it}$	Sales (revenues) of company i in year t
$EQUT_{it}$	Book value of equity (shareholders'/stockholders' equity) of company i in year t
TA_{it}	Book value of total assets of company i in year t
$EQUTRA_{it}$	Equity-to-assets ratio of company i in year t
FA_{it}	Fixed assets of company i in year t
$FATO_{it}$	Fixed assets turnover ratio = $SALES_{it}/FA_{it}$
$FCFE_{it}$	Free cash flow to equity of company i in year t
$FCFF_{it}$	Free cash flow to the firm of company i in year t
CFO_{it}	Operating cash flow (cash flow from operating activities) of company i in year t
ROE_{it}	Return on equity of company i in year t = $NI_{it}/EQUT_{i(t-1)}$
DIV_{it}	Dividends of company i paid for the year t

Stock price variables

Note: The stock price variables are measured on the last trading day in March of the year following the respective financial year. This is consistent with prior research. It is done to account for the typical financial statement disclosure lag of about three months after the balance sheet date. Thus, the aim of this proceeding is to measure stock price variables at about the same time at which investors receive the financial statement information. The stock price variables are as follows:

$PRICE_{it}$	Stock price of company *i* on the last trading day in March of year *t*+1
EV_{it}	Enterprise value of company *i* in year *t*; defined as: $PRICE_{it}$+book value of debt of company *i* in year *t*

[175] For a general overview of similarities and differences between IFRS and US-GAAP, see PriceWaterhouseCoopers (2004).

$BETA_{it}$	Historical beta-factor of company *i* on the last trading day in March of the year *t*+1; calculated as the slope coefficient of a linear regression of single stock returns on stock index returns; calculations are based on daily returns for a period of 250 trading days; the DAFOX has been used as the benchmark index.

Dummy variables

$DY99_t$ - $DY03_t$	Yearly dummy variables for the years 1999 through 2003
$D1_i$	Dummy variable for the competitiveness of the industry $D1_i$ =1 if industry is competitive, 0 otherwise
$D2_{it}$	Dummy variable for the asset heaviness of companies $D2_{it}$ =1 if company is asset heavy, 0 otherwise
$D3_{it}$	Dummy variable for the creditworthiness of companies $D3_{it}$ =1 if company has a high creditworthiness, 0 otherwise
$D4_{it}$	Dummy variable for the earnings adjusted creditworthiness of companies $D4_{it}$=1 if company has a high earnings adjusted creditworthy ness, 0 otherwise.
$NINEG_{it}$	Dummy variable for negative NI_{it} $NINEG_{it}$ =1 if company has negative NI, 0 otherwise

Other variables

CRW_{it}	Creditworthiness (i.e. the Creditreform Rating Index) of company *i* in year *t*
$ACRW_{it}$	Earnings adjusted creditworthiness of company *i* in year *t*
GDP_t	Real GDP, Germany, for the year *t*
$GDPEXPECT_t$	GDP expectations, Germany, measured as the value of the "Indicator of Economic Sentiment" published by the ZEW in March of year *t*+1

5.2.3 Hypotheses and Econometric Methodology

5.2.3.1 Hypotheses

To structure the empirical analysis of value relevance, a set of testable hypotheses is formulated building on the theoretical parts of this thesis (chapters 3 and 4). The first two hypotheses correspond to the role of cash flows in CCV. This has been discussed in section 3.2.1.4.

Hypothesis 1.1 *Net income is more value relevant than operating cash flow*

Hypothesis 1.2 *Free cash flow to equity is less value relevant than each of the following accounting figures: net income, book value of equity and operating cash flow*

The next hypothesis relates to enterprise level CCV. The focus is on the role of sales multiples in business valuation. The hypothesis is derived from the theoretical presentation in section 3.2.1.2.

Hypothesis 2 *Sales have lower value relevance than each of the following accounting figures: EBIT, EBITDA*

A turn to section 4.3.2 (derivation of a two-factor model based on book value and earnings) is necessary to find the theoretical basis for the following hypothesis:

Hypothesis 3 *The relationship between the PB ratio and the return on equity is non-linear (i.e. rather convex)*

The next group of hypotheses also focuses on the joint value relevance of earnings and book value. This issue has been extensively discussed in section 3.2.1.3.

Hypothesis 4.1 *The joint value relevance of earnings and book value is higher than the value relevance of each of these two variables alone*

Hypothesis 4.2 *The value relevance of earnings is reduced and the value relevance of book value is increased for firms that operate in highly competitive industries compared to firms that do not*

Hypothesis 4.3 *The value relevance of earnings is reduced and the value relevance of book value is increased for firms that are asset heavy compared to firms that are not*

Hypothesis 4.4 *The value relevance of earnings is reduced and the value relevance of book value is increased for firms that have a high relative creditworthiness compared to firms that have a poor relative creditworthiness*

The next two hypotheses are related to some sort of "extended" value relevance. They are a logical consequence from the theoretical discussion in section 3.2.1.3. These hypotheses aim at explaining the height of multiples under certain circumstances.[176]

---

176 Note: The relative value is the value for a given set of accounting variables; it is closely related to the multiples.

Hypothesis 5.1 *A company has a lower relative value if it operates in highly competitive industries than if it operates in non-competitive industries – assuming accounting figures are the same in both cases*

Hypothesis 5.2 *A company has a lower relative value if it has a low degree of creditworthiness than if it has a high degree of creditworthiness – assuming accounting figures are the same in both cases*

While Hypothesis 5.2 is intuitively reasonable (a company with a high creditworthiness supposably has lower cost of capital due to the lower financial risk, which reduces the capitalisation rate in direct valuation approaches and finally leads to an increase in stock price), Hypothesis 5.1 requires two short explanatory notes. First, companies in very competitive industries are expected to have a relatively low persistence of abnormal earnings. This forces earnings to decline over time and supposably results in lower stock prices. Second, underperforming companies recover worse and slower in highly competitive industries. This presumably drives the future cash flow stream to generally be lower, and finally forces stock prices to fall short of prices of companies in non-competitive industries.

5.2.3.2 Measurement Issues

In order to correctly specify the dummy variables $D1_i$, $D2_{it}$, $D3_{it}$ and $D4_{it}$, some rules of measurement have to be framed. In fact, it is important to establish criteria based on which companies can be classified into "asset heavy" or "not asset heavy", and into "highly creditworthy" or "not highly creditworthy". Additionally, criteria must be defined for which industries can be regarded as competitive in the sense of *Porter*.

Dummy variable D1: competitiveness of industries

Following Porter's five forces analysis, a competitive industry is characterised by low entry barriers, high rivalry among existing competitors, high buyer power, high supplier power and a serious threat of substitute products (see Porter, 1980: 3-29; Porter, 1998: 21-34). In order to classify industries into either competitive or non-competitive, a scoring model is applied.[177] This model ranks each industry according to its weakness (very weak companies are ranked first) for each of the five forces, and subsequently adds up these ranks to generate an overall score for each industry. According to that score, the top six industries (the top third of the industries under consideration) are denoted "non-competitive", the middle six industries are denoted "neutral" and the lower six industries (the bottom third of the industries under consideration) are denoted "competitive".

---

177 For the present analysis, industries are defined as groups of certain NACE classes. The classification is largely identical with that of Sofka and Schmidt (2004). For a complete list of all industries and NACE classes included, see Appendix 7.3. However, due to a lack of data, only 18 of 23 industries are analysed in the Porter five-forces examination.

Barriers to entry are measured as the number of business start-ups (average value for the years 2002/2003) for each existing company in the industry. A low number indicates high barriers to entry, while a high number indicates low barriers to entry. The industry with the lowest number is ranked first. Data are obtained from the ZEW Start-Up Panel.

Rivalry among competitors is measured by the Herfindahl index for each industry. A high index indicates a low degree of rivalry, while a low index indicates a high degree of rivalry (see Solnik and McLeavy, 2004: 265-266). The industry with the highest number is ranked first. Data are obtained by the Expert Opinion of the Monopoly Commission (Hauptgutachten der Monopolkommission) 2001. Some additional calculations were necessary. Herfindahl indexes in this analysis are bound between 0 and 1,000. The calculation scheme is as follows:

$$\text{Herfindahl - Index} = 1,000 \cdot \sum_{i=1}^{m} \left(SALES_{firm\, i} / SALES_{industry} \right)^2$$

with *n* being the number of firms in the industry under consideration.

Buyer power is measured by the percentage of sales that is absorbed by the three biggest customer industries. The higher this percentage is, the more concentrated are the buyers and the more power lies in their hands. The industry with the lowest number is ranked first. Data are obtained from the input-output tables published by DESTATIS in 1998. It is assumed that these industry level relations remain quite stable over time.

Supplier power is measured by the percentage of sales that is put into the production process by the three biggest supplier industries. The higher this percentage is, the more concentrated are the suppliers and the more power lies in their hands. As in the assessment of buyer power, the industry with the lowest number is ranked first. Data are obtained from the input-output tables published by DESTATIS in 1998. It is assumed that these industry level relations remain quite stable over time.

Threat of substitutes is measured as the fraction of industry sales that is generated by product innovations (average value for the years 2002/2003). The higher this percentage is, the higher is the threat of substitute products. Consequently, the industry with the lowest number is ranked first. Data are obtained from the Mannheim Innovation Panel (MIP).

Table 4 shows the values of the measures and the ranks for each force and industry.[178] Additionally, the overall score, the overall rank, and the classification of each industry is indicated. Industry numbers correspond to the list in Appendix 7.3. The dummy rule is:

$\mathbf{D1_i = 1}$ if an industry is competitive (i.e. industries with overall ranks 13-18)

$\mathbf{D1_i = 0}$ otherwise (i.e. industries with overall ranks 1-12)

---

178 See Appendix 7.4 for some descriptive statistics of the sample by industry classification.

Table 4. Porter's five forces analysis

Industry No. [a, b]	Porter's Categories										**Overall Score** [h]	**Overall Rank**	Classification of Industry
	Barriers to Entry [c]	*Rank*	Rivalry among Competitors [d]	*Rank*	Buyer Power [e]	*Rank*	Supplier Power [f]	*Rank*	Threat of Substitutes [g]	*Rank*			
19	0.250	*11*	0.1	*18*	0.317	*5*	0.658	*17*	0.065	*12*	**63**	**18**	competitive
13	0.537	*17*	18.1	*16*	0.332	*6*	0.661	*18*	0.023	*2*	**59**	**17**	competitive
11	0.350	*15*	68.7	*7*	0.686	*13*	0.438	*3*	0.127	*17*	**55**	**16**	competitive
2	0.270	*12*	54.2	*9*	0.739	*16*	0.449	*4*	0.096	*14*	**55**	**15**	competitive
17	0.779	*18*	212.5	*2*	0.301	*3*	0.64	*14*	0.123	*16*	**53**	**14**	competitive
1	0.203	*9*	159.8	*4*	0.946	*18*	0.504	*7*	0.121	*15*	**53**	**13**	competitive
6	0.173	*6*	127.2	*5*	0.503	*10*	0.537	*11*	0.144	*18*	**50**	**12**	neutral
18	0.318	*14*	41.5	*12*	0.283	*2*	0.523	*10*	0.065	*11*	**49**	**11**	neutral
10	0.162	*4*	60.2	*8*	0.690	*14*	0.58	*13*	0.058	*10*	**49**	**10**	neutral
5	0.087	*1*	24	*15*	0.499	*9*	0.654	*16*	0.049	*8*	**49**	**9**	neutral
12	0.170	*5*	70.9	*6*	0.748	*17*	0.649	*15*	0.031	*4*	**47**	**8**	neutral
3	0.117	*2*	47	*11*	0.663	*12*	0.515	*9*	0.071	*13*	**47**	**7**	neutral
9	0.206	*10*	34.2	*14*	0.509	*11*	0.489	*6*	0.035	*5*	**46**	**6**	Not comp.

14	0.358	*16*	8.9	*17*	0.305	*4*	0.474	*5*	0.015	*1*	**43**	**5**	Not comp.
7	0.176	*7*	47.6	*10*	0.709	*15*	0.405	*1*	0.044	*6*	**39**	**4**	Not comp.
15	0.313	*13*	504.0	*1*	0.380	*7*	0.568	*12*	0.027	*3*	**36**	**3**	Not comp.
8	0.189	*8*	34.5	*13*	0.240	*1*	0.408	*2*	0.045	*7*	**31**	**2**	Not comp.
4	0.128	*3*	289.6	*3*	0.426	*8*	0.51	*8*	0.050	*9*	**31**	**1**	Not comp.

[a] Industry number according to the classification in Appendix 7.3.

[b] Missing: Industry No. 16 (Financial Intermediation), 20 (Real Estate Activities and Renting), 21 (Sporting and other Business Activities), 22 (Utilities) and 23 (Construction). Reason: No data available.

[c] Measured by the number of business start-ups (average 2002/2003) per each existing company in the industry, Source: ZEW Start-Up Panel.

[d] Measured by the Herfindahl-Indexes of each industry, Source: Appendix to the Expert Opinion of the Monopoly Commission (Hauptgutachten der Monopolkommission) 2001; own calculations.

[e] Measured by the percentage of sales that is absorbed by the three biggest customer industries, Sources: Input-Output Tables 1998, DESTATIS.

[f] Measured by the percentage of sales that is put in by the three biggest supplier industries, Sources: Input-Output Tables 1998, DESTATIS.

[g] Measured as the fraction of sales that is generated by product innovations (average 2002/2003), Source: Mannheim Innovation Panel (MIP)

[h] Sum of category ranks.

Dummy Variable D2: Asset Heaviness of Companies

Asset heavy (capital intensive) companies are characterised by relatively high fixed costs and low asset turnover ratios (see White et al., 1997: 189-190). Consequently, for the present analysis asset heaviness is measured by the fixed asset turnover ratio (FATO). This ratio is as follows:

$$FATO_{it} = SALES_{it} / FA_{it}$$

The cut-off value for dividing into "asset heavy" and "not asset heavy" companies is set at 2.5. Using this cut-off value ensures that both resulting groups of companies have approximately the same size.[179] Thus, the dummy rule is:

$D2_{it}$ = 1 if a company has $FATO_{it} \leq 2.5$ (i.e. the company is asset heavy)
$D2_{it}$ = 0 if a company has $FATO_{it} > 2.5$ (i.e. the company is not asset heavy)

Dummy Variable D3: Creditworthiness of Companies

The construction of the creditworthiness dummy variable is based on the rating provided by Creditreform. Creditreform uses different information data for its rating. These are, in particular, liquidity and financial risks as well as structural risks like firm age, firm size and productivity, along with certain "soft factors" like payment history, volume of orders, firm development, management quality etc. On the basis of the individual facts, Creditreform determines a rating index (i.e. a score, denoted CRW) ranging from 100 points to the maximum of 600 points. The worst firms receive 600 points and the best ones have 100 points. For their customers, the rating agency constructs a six-class rating, which is presented in Table 5.[180]

Table 5. The rating by Creditreform

Original Rating Classes of Creditreform	Creditworthiness	Rating Index
1	very good	[100-130)
2	good	[130-200)
3	average	[200-300)
4	weak	[300-400)
5	insufficient	[400-500)
6	turn away business connection	[500-600]

Source: Czarnitzki and Kraft; 2004: 6.

For the present analysis, "highly creditworthy" is defined as rating class 1 (i.e. a rating index between 100 and 130). Thus, the dummy rule is:

179 See Appendix 7.4 for some descriptive statistics of the sample by asset heaviness of companies.

180 See Czarnitzki and Kraft (2004). See Appendix 7.4 for some descriptive statistics of the sample by creditworthiness of companies.

$D3_{it}$ = 1 if a company has $CRW_{it} < 130$ (i.e. high creditworthiness)
$D3_{it}$ = 0 if a company has $CRW_{it} \geq 130$ (i.e. no high creditworthiness)

Dummy Variable D4: Earnings Adjusted Creditworthiness of Companies
A major problem in the empirical analysis is that creditworthiness is supposed to influence earnings' value relevance in two conflicting ways. The first one is based on the "financing of investments" theory. It states that companies with high creditworthiness do not have debt financing problems and, thus, do not solely rely on plowing back earnings in order to finance future investments. This leads to the conjecture that high creditworthiness has a negative influence on earnings' value relevance.[181] However, contrary to that, a high creditworthiness is typically associated with strong balance sheets and healthy income statements. As a consequence, highly creditworthy companies typically have high positive earnings, and high positive earnings, in turn, are associated with a high relative value relevance of earnings (see Burgstahler and Dichev, 1997; Barth et al., 1998; Collins et al., 1999).

The net effect of creditworthiness has been captured by dummy variable D3. However, the goal is here to examine the validity of the first of the two theories (the "financing of investment" theory). For this analysis, some adjustments to creditworthiness are necessary. If the predicted influence exists, then a highly creditworthy company has a relatively lower value relevance of earnings than a company with bad access to the debt market, assuming both companies have the same level of earnings. To put it differently, creditworthiness has a negative effect on earnings value relevance after controlling for the level of earnings. To assess this relationship, the impact of current earnings on creditworthiness has to be filtered out. The filtering technique proceeds in two steps. In the first step, creditworthiness is regressed on earnings scaled by price[182] with the intercept suppressed.[183] The residuals of this regression capture that part of creditworthiness that is not explained by earnings. Thus, they are a measure of earnings adjusted creditworthiness. In the second step, the residuals are defined as a new variable.[184] This variable also has – just like the creditworthiness itself – an ordinal scale of measurement.

181 See section 3.2.1.3.

182 The scaling is done in order to normalise earnings, and to account for the fact that size has no direct impact on creditworthiness.

183 Note: Even if the depend variable has an ordinal scale of measurement, this regression technique is valid and still yields the best linear predictor as long as the dependent variable is continuous. In the present case, creditworthiness can be seen as a quasi-continuous variable. However, in order to check for robustness, the whole earnings adjustment process was repeated using a Box-Cox transformation of the variable CRW as dependent variable (see Appendix 7.6 for a detailed presentation of the proceeding and of the results of this transformation approach). Since the results of both estimations are virtually identical and the implications are qualitatively the same, it is concluded that the earnings adjustment technique applied here is robust.

184 This proceeding is similar to the proceeding of Schröder (2003: 26).

Table 6. Extracting the impact of earnings out of creditworthiness

Regression model

$$CRW_{it} = b_1 \cdot \left(\frac{NI_{it}}{PRICE_{it}} \right) + \varepsilon_{it}$$

Regression results

Variable[a]	Coefficient	p-value	Uncentred R^2
$\frac{NI}{PRICE}$	-37.42	0.00	0.150

Thus, earnings adjusted creditworthiness ACRW is calculated as follows:

$$ACRW_{it} = \hat{\varepsilon}it = CRW_{it} - \hat{b_1} \cdot \left(\frac{NI_{it}}{PRICE_{it}} \right) = CRW_{it} + 37.42 \cdot \left(\frac{NI_{it}}{PRICE_{it}} \right)$$

[a] Variables are as defined in section 5.2.1.

With that new measure of earnings adjusted creditworthiness (denoted ACRW) the dummy variable D4 can be specified. In order to maintain consistency with the dummy variable D3, the cut-off rate to divide the sample into "highly earnings adjusted creditworthy" and "not highly earnings adjusted creditworthy" companies is set such that the percentage of companies in both groups remain the same as under dummy D3. Consequently, the new cut-off rate is at ACRW=123. Thus, the dummy rule is:

$D4_{it}$= 1 if a company has a $ACRW_{it}<123$ (i.e. high earnings adjusted creditworthiness),

$D4_{it}$= 0 if a company has a $ACRW_{it}\geq 123$ (i.e. no high earnings adjusted creditworthiness).

5.2.3.3 General Research Design

Hypotheses 1.1, 1.2 and 2 are all tested with the same set of estimation equations. According to other studies investigating the value relevance of single variables, the following model specifications are adopted:

$$Y_{it} = b_0 + b_1 \cdot X_{it} + \varepsilon_{it} \tag{5.1}$$

$$Y_{it} = b_1 \cdot X_{it} + \varepsilon_{it} \tag{5.2}$$

$$Y_{it} = b_0 + b_1 \cdot X_{1,it} + b_2 \cdot X_{2,it} + \varepsilon_{it} \tag{5.3}$$

where Y is a replacement character for the dependent variables *PRICE* or *EV*, and X is a replacement character for the independent variables *NI*, *CFO*, *EQUT* and *FCFE* resp. *SALES*, *EBIT*, *EBITDA*, *TA* and *FCFF*. The estimation equations consider the different requirements of equity valuation and enterprise valuation, i.e. all equity figures are related to *PRICE* while all enterprise figures are related to *EV*.

Equation 5.1 is estimated for two different samples: In the first step it is estimated for the whole sample to examine the real value relevance of accounting variables. In the second step, it is estimated only for observations with positive X_{it} in order to better assess the value relevance in the context of CCV since in classical single-factor models only positive reference variables are applied. Equation 5.2 is even more focused on single-factor CCV since it suppresses the intercept and therefore considers that valuation multiples typically pass the origin. Equation 5.3 is estimated to show the value relevance of single variables if applied jointly with other variables in CCV. This is an important issue for appraisers who plan to apply multi-model approaches.

Value relevance is measured by the common (centred) R^2 in Equation 5.1 for both the whole sample and the only-positive sample. Since the centred R^2 produces a biased assessment of the model fit if the intercept is suppressed, value relevance in Equation 5.2 is measured by the uncentred R^2. The uncentred R^2 is calculated for a regression with n observations as follows:

$$\text{uncentred } R^2 = \sum_{i=1}^{n} \left(\hat{Y}_i\right)^2 \Big/ \sum_{i=1}^{n} \left(Y_i\right)^2$$

with $\hat{Y}_i$ being the value that is predicted by the regression model.

It is important to note that the results of the uncentred R^2 are *not* comparable with the results of the centred R^2. However, a comparison of uncentred R^2 with another uncentred R^2 is, of course, possible.

Value relevance in Equation 5.3 is measured by the incremental R^2. The incremental R^2 gauges the contribution of one variable to the R^2 of the whole regression model. It is calculated for variable X_1 given the variable X_2 as follows (see Theil, 1971: 168-171; Ota, 2001: 9):

$$\text{incremental } R^2_{X_1} = \text{adj.}R^2_{X_1,X_2} - R^2_{X_2}$$

i.e. the incremental R^2 for variable X_1 given X_2 is the difference between the adjusted R^2 of the multiple regression (with X_1 and X_2 as independent variables) and the R^2 of a regression excluding X_1 from the set of independent variables.

Tests of Hypothesis 3 are also based on the estimation Equation 5.1.[185] For the analysis the whole sample is divided into three sub-samples according to the value of the independent variable, e.g. one sub-sample contains observations for highly negative *ROE* up to slightly negative *ROE*, another sub-sample ranges from slightly negative to slightly positive *ROE* and a third sub-sample covers all observations from slightly positive to highly positive *ROE*. Sub-sample sizes are chosen such that the number of observations is about the same in each group. In the next step, the slope coefficients of each piece-wise regression are compared to investigate whether there are significant differences. The existence of such differences would be an indicator of non-linearities in the relationships.[186]

Hypothesis 4.1 is tested based on Equations 5.1 and 5.3, with *NI* and *EQUT* being the variables X_1 and X_2. Value relevance is measured by adjusted R^2. Hypotheses 4.2, 4.3 and 4.4 are tested in basically two ways:

(1) Each of these hypotheses is tested separately. This is done with a regression based on Equation 5.3, with *NI* and *EQUT* being the variables X_1 and X_2, for each of the two groups (the one for which D*X*=0 holds and the one for which D*X*=1 holds).[187] Then, the coefficients are tested for significant differences based on the assumption that the covariance between the same coefficients for each group is zero. Additionally, the incremental R^2 of each variable and group is used as a measure of value relevance. This approach allows the two groups having different intercepts and different time dummies.

Additionally, a regression is run that explicitly includes the dummy variable for each coefficient. In this regression, an intercept dummy is also included in order to assess the value relevance of a change from D*X*=0 to D*X*=1 in its own right:

$$PRICE_{it} = b_0 + b_1 \cdot DX_{it} + b_2 \cdot EQUT_{it} + b_3 \cdot DX_{it} \cdot EQUT_{it} + \\ + b_4 \cdot NI_{it} + b_5 \cdot DX_{it} \cdot NI_{it} + \varepsilon_{it} \qquad (5.4)$$

with D*X* being a replacement character for the dummy variables D1, D2, D3 and D4.[188]

In this model the slope coefficient b_4 represents the value relevance of earnings for the set of companies for which D*X*=0 holds. Similarly, the slope coefficient b_2 represents the value relevance of book value for that group of companies. The

185 Now the dependent variable is the PB ratio and the independent variable is *ROE*.

186 The proceeding is similar to the methodology in Burgstahler and Dichev (1997).

187 Here, D*X* is a replacement character for the dummy variables D1, D2, D3 and D4.

188 Note: For the variable D1 the subscript is only *i*, not *it*. Thus, D1 does not vary over time.

market's total response to earnings for the group of companies for which DX=1 holds is calculated by the sum of the coefficients b_4 and b_5 (assuming both are significant). Likewise, adding the (significant) coefficients b_2 and b_3 yields the value relevance of book value for the (DX=1) group of companies. Thus, the coefficients b_3 and b_5 indicate how value relevance varies between the (DX=0) companies and the (DX=1) companies. Note that the focus here is on change in value of coefficients rather than on R^2 because the aim is to determine differences in the *relative* value relevance of book value and earnings. The basic assumption of this approach is a common variance of coefficients for both groups. Thus, the constant term and the time dummies are estimated commonly, too.

(2) The second way is to test all hypotheses commonly in one complete regression model. This model has the following form:

$$\begin{aligned} PRICE_{it} = {} & b_0 + b_1 \cdot D1_i + b_2 \cdot D2_{it} + b_3 \cdot D3_{it} + b_4 \cdot D4_{it} + \\ & + b_5 \cdot EQUT_{it} + b_6 \cdot NI_{it} + \\ & + b_7 \cdot D1_i \cdot NI_{it} + b_8 \cdot D2_{it} \cdot NI_{it} + b_9 \cdot D3_{it} \cdot NI_{it} + b_{10} \cdot D4_{it} \cdot NI_{it} + \\ & + b_{11} \cdot D1_i \cdot EQUT_{it} + b_{12} \cdot D2_{it} \cdot EQUT_{it} + \\ & + b_{13} \cdot D3_{it} \cdot EQUT_{it} + b_{14} \cdot D4_{it} \cdot EQUT_{it} + \varepsilon_{it} \end{aligned} \tag{5.5}$$

The interpretation of Equation 5.5 is similar to the interpretation of Equation 5.4 with the only difference that now all dummy terms are part of one large regression model. This model helps to determine whether the results from the previous estimations are robust. To further check for robustness of this large model, a second version of Equation 5.5 is estimated. This version differs from the first in that a set of control variables is also included in the estimation.

5.2.3.4 Research Design of the Matching Estimator Approach

The problem of a selection bias associated with Hypotheses 5.1 and 5.2

The research questions of Hypotheses 5.1 and 5.2 are the following: (1) Does a company that operates in a non-competitive (or neutral) industry trade at a higher price than a company with the same financial figures that operates in a highly competitive industry? (2) Does a company that has a high degree of creditworthiness trade at a higher price than a company with the same financial figures that has a low creditworthiness? Or differently for both: Does industry membership (or creditworthiness) have an impact on the height of multiples?

Note: Since both questions are addressed by the same research approach, the introduction of the terms *treatment group* and *control group* will clearly make the further explanations more understandable. As regards the first research question, the group of companies that operate in competitive industries is called the treatment group (i.e. the companies are characterised by the treatment "competitive industry") while the other companies are called the control group. Likewise, in the second research case the group of companies that have a high creditworthiness is called the treatment group, while the remaining companies are called the *control group*.

Of course, the two research questions of Hypotheses 5.1 and 5.2 cannot be addressed by pure observation. This is the case because companies' industry membership is exclusive and companies either have a high creditworthiness or they do not. However, what is even more problematic is that a direct comparison of the two sub-samples[189] – treatment group and control group – cannot be performed reasonably because there is a risk of a *sample selection bias*. To put it more precisely, companies that face strong industry forces must run totally different business strategies than companies that face weak industry forces. Similarly, companies with a high degree of creditworthiness face different financial challenges than companies with a low degree of creditworthiness. This has an impact on the accounting figures as well; i.e. two companies (one from the treatment group and one from the control group) that are quite identical in any qualitative criteria other than industry classification or creditworthiness might still have totally different accounting figures. This finally implicates that on average the two groups do by no means have the same financial statements.[190] Thus, neither industry membership nor creditworthiness can be reasonably interpreted as the result of random process.

While this is not a big problem in a pure assessment of value relevance, it becomes a major problem when assessing the differences in relative values. Why? In pure value relevance research the concern is about how *different* accounting figures and (potentially) *different* prices interact, while now the focus is on how *identical* accounting figures would (potentially) translate in *different* values given a *different* industry environment or a *different* creditworthiness. However, since the identity in accounting figures for both groups cannot be assumed, the problem of a potential selection bias is of special significance now.

To conclude: Assessing the impact of industry membership and creditworthiness – called *treatment* hereafter – on relative values, based on a comparison of the treatment groups and the control groups, without correction for that selection bias, may generate misleading results.

Econometric methodology of the matching estimator approach

There are different strategies to correct such a selection bias.[191] Most of them, however, are not appropriate for the present case. For example, the difference-in-difference method requires data before and after treatment (but there are only rare – if at all – data about companies that switch from one industry classification to the other group; the same is true for the classification of creditworthiness) and the so-called instrumental variables estimators require valid instruments (but it is very difficult, if not impossible, to find valid candidates that can serve as instruments in

189 This procedure has been applied in the previous analyses.

190 A look at the descriptive statistics by industry structure and by creditworthiness (Appendix 7.4) emphasizes the differences between both groups – not only in prices but also in accounting figures.

191 An overview is provided by Heckman et al. (1999).

this analysis). Fortunately, the third method – the matching estimator approach – does not have any of these unsatisfiable requirements. Moreover, it is not based on the assumption of any functional form of the outcome equation – i.e. of industry classification or creditworthiness – nor is an assumption necessary with respect to the shape of the distribution of the error terms of the outcome equation and the selection equation (see Czarnitzki et al., 2004: 9).

Matching estimators are a relatively young econometric methodology. They have been applied and discussed mainly in the field of labour market research (see Angrist, 1998; Dehejia and Wahba, 1999; Heckman et al., 1998; Lechner, 1999a; Lechner, 1999b; Lechner, 2000; Gerfin and Lechner, 2002 and Vandenberghe and Robin, 2004), but most recently also in the field of industrial economics (see Arnold and Hussinger, 2004; Czarnitzki et al., 2004). However, it has not been used very often in finance research thus far. Therefore, this study is somewhat unique in the way it goes about testing these two asset-pricing related hypotheses.

The principle of matching is the following: The starting situation is that two groups of observations are in different states. The control group is defined as being originally in state *m*, while the treatment group is defined as being in state *l*. The matching estimator now establishes a counterfactual situation for companies that are part of the control group, so that they are virtually in the state of the treatment group.[192] To put it more precisely, the matching estimator is the tool to create a sample of firms in *l* that is comparable to the sample of companies in *m*, where comparability relates to a set of a priori defined criteria. This set of criteria (also called covariates) is denoted *G*. The consequence of this matching is that – once the samples in states *m* and *l* are comparable with respect to *G* – remaining differences in the outcome between both groups can exclusively be assigned to the treatment (i.e. to industry classification or creditworthiness). This difference is called the *treatment effect*. In formal terms, the average treatment effect on the control group (companies in state *m*,) relative to state *l* can be expressed as (see Czarnitzki et al., 2004: 8, Lechner, 1999b: 4):

$$E\left[\alpha^{m,l}\right]=E\left[Y^{m}-Y^{l}\,|S=m\right]=E\left[Y^{m}\,|S=m\right]-E\left[Y^{l}\,|S=m\right] \tag{5.6}$$

Where $E\left[\alpha^{m,l}\right]$ denotes the average treatment effect, Y^{m} and Y^{l} denote the outcome in the different states and *S* indicates the factual state. Figure 31 summarises the effects of matching on the control group and the treatment group.

192 In most research a binary treatment is applied, i.e. the treatment variable can have two states. This is also done in the present case. However, in some research this approach has been extended to multiple state matching, see e.g. Gerfin and Lechner (2002), Czarnitzki et al. (2004).

Fig. 31. Effects of matching

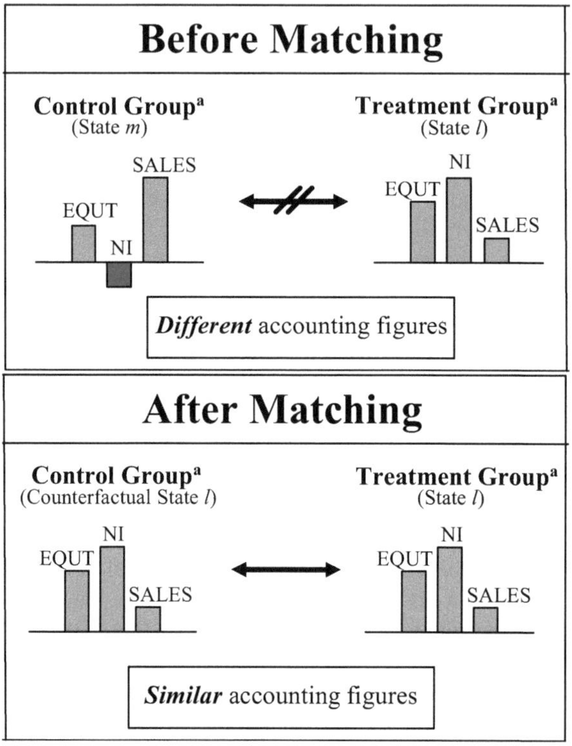

[a] The bars in the figures do not represent real values of accounting variables, they are just examples.

This approach requires the definition of the similarity criteria *G*. In the present analysis the variables *NI*, *EQUT*, *EQUTRA* and *SALES* should suffice to define comparability. The matching procedure itself proceeds as follows: First a *probit regression* of the probability of treatment as a function of the covariates *G* is run. In the second step, the resulting estimates are used to create a predicted probability of treatment for each observation (this probability is called the *propensity score*).[193] Then pairs of firms from each group can be identified that have the same probability of receiving treatments, i.e. balancing on the propensity score results in matched samples that should be similar in the covariates *NI*, *EQUT*, *EQUTRA* and *SALES*. In order to achieve valid matching results, it is important to apply a *common support*, i.e. to allow all firm year observations to participate in both states. The common support assumption can be accounted for by deleting all observa-

[193] The idea of determining the propensity scores via probit regressions goes back to Maddala (1983).

tions with probabilities larger than the smallest maximum and smaller than the largest minimum of both groups (see Czarnitzki et al., 2004: 10).

However, even within common support, the probability of observing two companies with exactly the same propensity score is about zero in practice. To overcome this difficulty, several solutions are offered in literature (see Vandenberghe and Robin, 2004). One of them is the *nearest neighbour matching approach*; it consists of an algorithm that matches each company from the control group with the treatment group according to the nearest propensity score. Thus, the resulting match minimises the bias between the treatment and comparison groups. However, applying this method is probably associated with a disregard for important information since not necessarily all observations from the treatment group are matched. This is especially a problem if the number of observations in the treatment group exceeds the number of observations in the control group, which is the case in the present examination where the competitive industry group consists of 438 observations, while the control group only consists of 372 observations.[194] As a consequence, another matching algorithm is applied here: the *Epanechnikov kernel matching*. In Epanechnikov kernel matching, all members of the non-treatment group are used in order to build a match for each member of the treatment group (of course, the contribution of those for whom the match is poor is rather small). To put it differently, the kernel is a function that weights the contribution of each control group member according to distance of propensity scores. Exact matches get a large weight, and poor matches get a small weight.

After this matching procedure is applied, it is possible to determine the difference in the variable *PRICE* between the two groups that can exclusively be assigned to industry classification resp. creditworthiness. Thus, in the ideal case, all differences in accounting figures are filtered out and the selection bias is corrected. Figure 32 summarises the matching protocol.[195]

---

194 In the examination of the impact of creditworthiness, this is less of concern since the treatment group contains 101 observations while there are 596 observations in the control group.

195 The matching is done with the assistance of the *psmatch2* procedure by Leuven and Sianesi (2003).

Fig. 32. Matching protocol

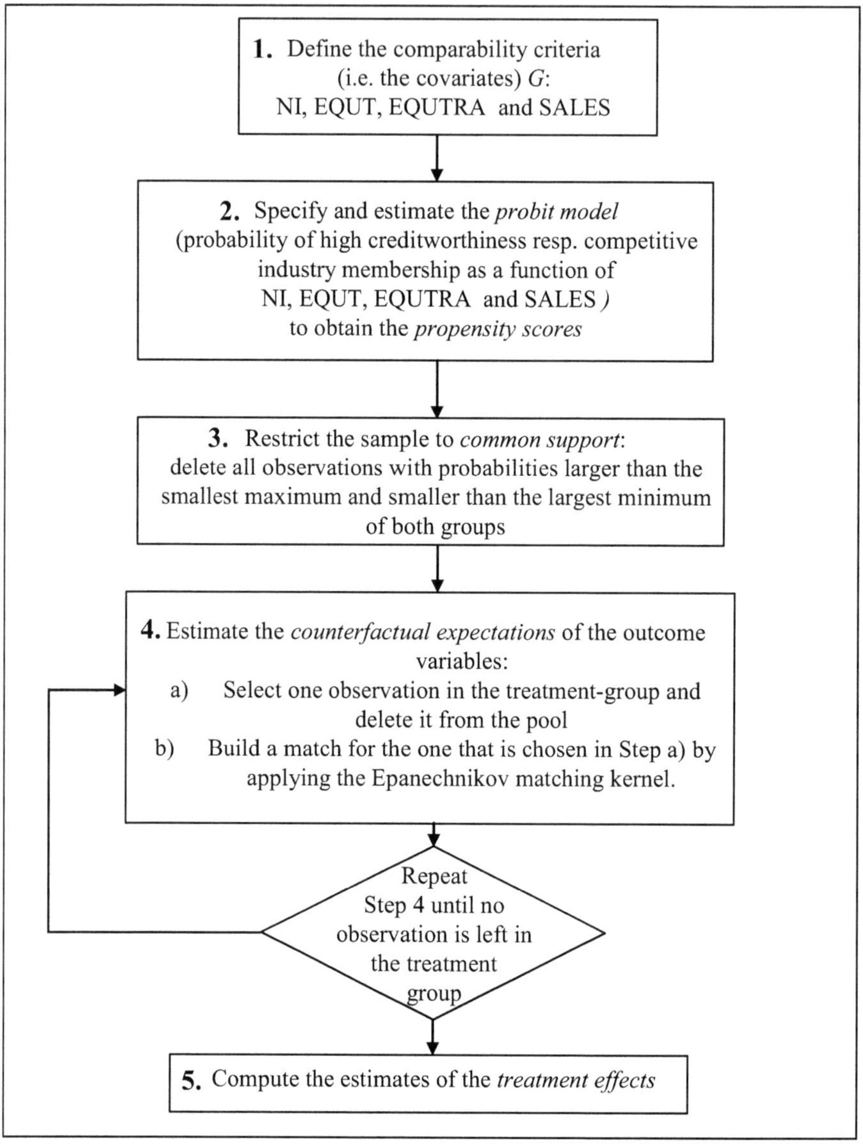

Source: Own figure according to Czarnitzki et al., 2004: 11, Lechner, 1999b: 16.

5.2.4 Results

5.2.4.1 Descriptive Statistics

Table 7 shows some basic statistics on the sample.[196] Especially two points are striking here. First, the mean of the variable *NI* is slightly negative, which is the consequence from the many negative-earnings-companies in the sample. One of the main reasons for this is that the sample period (1998-2003) covers the rise and the fall of the "Neuer Markt", a market segment for high growth companies. Even if several companies in this segment were not included in the sample because they ceased to be listed or were taken over before 2003, these mostly low profitable firms still have some influence on the sample. Second, the standard deviations of most of the variables are quite high. This wide dispersion is less an indicator for heterogeneity of the sample in the cross section (this is accounted for by focusing on companies that publish financial statements according to IAS/IFRS or US-GAAP) but rather a result of the macroeconomic volatility and the price level changes on the stock market during the sample period. Figure 33 illustrates the development of GDP growth rates in Germany and the DAX, the major German stock market index between 1998 and 2003.

Table 7. Basic descriptive statistics

Variables[a]	Number of observations	Mean	Standard deviation	Negative values, %
PRICE	928	14.53	17.79	0.0
EV	928	37.11	59.40	0.0
NI	928	-0.02	3.33	47.4
EBIT	928	0.61	4.51	45.2
EBITDA	928	2.69	7.27	29.0
SALES	928	33.31	75.02	0.0
FCFE	928	4.55	13.46	22.0
FCFF	928	4.15	11.01	23.4
CFO	928	1.81	5.83	34.3
EQUT	928	10.07	15.54	1.7
TA	928	32.65	63.07	0.0

[a] Variables are as defined in section 5.2.1.

196 Sample statistics by creditworthiness, by industry structure and by asset heaviness can be found in Appendix 7.4.

Fig. 33. Development of the stock market and the economy in Germany during the sample period

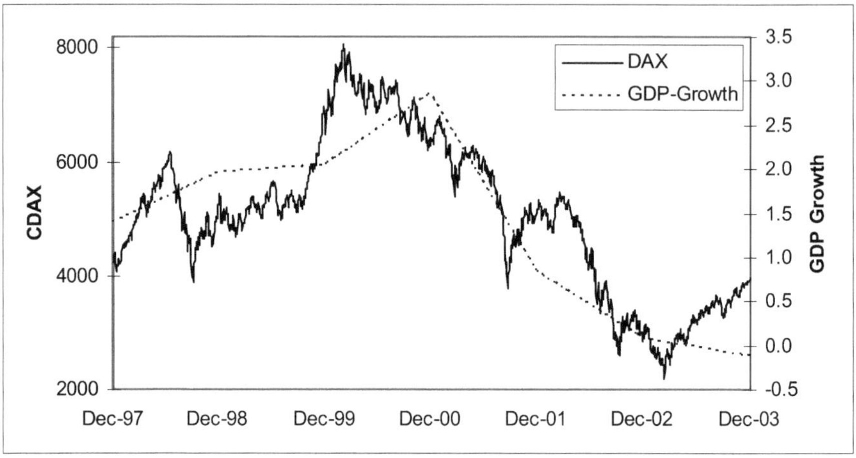

Source: DESTATIS, Datastream.

Pearson and Spearman correlations are reported in Table 8. Pearson correlation coefficients are a measure of linear association between two variables of interest. The Spearman coefficient is the non-parametric counterpart to the Pearson coefficient. It is regarded as the safer measure if the association between the two variables under consideration is rather non-linear. All correlations in the table are significant at the two tailed 1%-level. For the fields that are labelled "X", no correlations are calculated because this would not make any economic sense. Correlations between *PRICE* and other variables can only be interpreted reasonably if these variables are determined on an equity basis. Similarly, correlations between *EV* and other variables can only be interpreted reasonably if these variables are determined on an enterprise basis. It is important to note that the correlations between *PRICE* and accounting variables cannot be compared with the correlations between *EV* and the accounting variables because of technical reasons. More details about the problems of a potential upward bias in the *EV*-correlations can be found in section 5.2.4.2 and in Appendix 7.2.

The Pearson coefficients are quite high in some cases which might indicate that there is a risk of *multicollinearity*. Multicollinearity describes the phenomenon that in a regression model two or more independent variables are highly correlated so that a reasonable interpretation of regression outputs is difficult. In general, the level at which multicollinearity becomes a problem is set at an absolute correlation value of 0.5 or higher.[197] During the further examinations, problems might arise in

---

197 See DeFusco et al. (2001: 459). Sometimes this multicollinearity level is even set at 0.7. However, it is not possible to rule out the existence of multicollinearity only based on pairwise correlations because there might be linear combinations of independent variables that are highly correlated.

the regression model that uses *EQUT* and *NI* as independent variables (Equations 5.3, 5.4 and 5.5). However, for the further analysis, it is assumed that multicollinearity is not of concern for especially two reasons. First, Pearson correlation is below 0.5 and Spearman correlation is only slightly above 0.5. This still seems to be low enough to gauge the absence of multicollinearity. Second, variants of this model have been extensively estimated in the past and the results were – as far as the author knows – never called into question due to multicollinearity. Whether the inclusion of control variables causes multicollinearity problems will be analysed by a comparison of t-values and F-statistics for the respective regression models later in this analysis.

Table 8. Correlations

Variables[a, b]	PRICE	EV	NI	EBIT	EBITDA	SALES	FCFE	FCFF	CFO	EQUT	TA
PRICE	-	0.66	0.47	Xc	X	X	0.32	X	0.44	0.51	X
EV	0.86	-	X	0.67	0.81	0.86	X	0.79	X	X	0.96
NI	0.69	X	-	0.92	0.31	0.46	0.24	0.43	0.53	0.47	0.45
EBIT	X	0.64	0.96	-	0.54	0.62	0.33	0.58	0.70	0.62	0.63
EBITDA	X	0.77	0.80	0.85	-	0.72	0.45	0.69	0.77	0.70	0.80
SALES	X	0.83	0.52	0.56	0.72	-	0.43	0.71	0.81	0.84	0.90
FCFE	0.53	X	0.44	0.46	0.62	0.59	-	0.85	0.49	0.55	0.57
FCFF	X	0.67	0.51	0.53	0.69	0.66	0.83	-	0.81	0.75	0.80
CFO	0.49	X	0.56	0.59	0.71	0.65	0.59	0.80	-	0.78	0.84
EQUT	0.75	X	0.55	0.56	0.66	0.71	0.59	0.60	0.51	-	0.91
TA	X	0.90	0.49	0.52	0.71	0.88	0.65	0.67	0.60	0.87	-

[a] Variables are as defined in section 5.2.1.
[b] Pearson (Spearman) correlations are reported above (below) the diagonal. All correlations are significant at the 1% level (two tailed).
[c] X: No correlations calculated because of lack of economic coherence.

However, multicollinearity is not the only possible problem in regression interpretation. *Conditional heteroscedasticity*, i.e. non-constant variance of the residuals that is conditional on the values of the independent variables in the regression (see DeFusco et al., 2001: 447), can also affect statistical inference. If conditional heteroscedasticity exists, the unadjusted standard errors will be too small and consequently the test statistics will be too high. From an economic point of view, the risk of conditional heteroscedasticity is high in this analysis, because of the high volatility in the macroeconomic environment during the sample period. In order to statistically assess whether the residuals are actually affected by conditional heteroscedasticity, a Cook-Weisberg test[198] is run for each regression. The test revealed that – as expected – the variance of the residuals is non-constant and correlated with the values of the independent variables in all regression models and, thus, conditional heteroscedasticity is present. As a consequence, to adequately address this problem, *White heteroscedasticity-consistent robust standard errors* (see White, 1980) were used for the tests of significance in all regression models.

Endogeneity, i.e. reverse causality, does not seem to be a major problem in this analysis since stock prices usually do not have a great impact on accounting figures of non-financial companies (see Meitner and Westerheide, 2005).

5.2.4.2 General Estimation Results and Robustness Checks

For all analyses, Pooled Ordinary Least Square (Pooled OLS) regression is used. Fixed-effects or random-effects models are not appropriate here because of the structure of the sample.[199] However, in order to account for a timely heterogeneity of the sample and to diminish the impact of potential macroeconomic shocks, a dummy variable for each year of observation is included into the analysis.[200] For convenience, the hypotheses are listed again before presenting the respective results.

Hypothesis 1.1 *Net income is more value relevant than operating cash flow*

Hypothesis 1.2 *Free cash flow to equity is less value relevant than each of the following accounting figures: net income, book value of equity and operating cash flow*

Hypothesis 2 *Sales have lower value relevance than each of the following accounting figures: EBIT, EBITDA*

Table 9 presents the results for Equation 5.1 and 5.2 regressions. Equation 5.1 is estimated twice. First, the full sample is used and second, only positive independent variables are applied. Equation 5.2 is estimated only for positive independent variables. Common (centred) R^2 is indicated for the two Equation 5.1 models, while uncentred R^2 is indicated for Equation 5.2. Again, centred R^2 and uncentred R^2 are not directly comparable. The numbers in bold letters represent the highest values for each class (equity and enterprise value variables) and estimation model. The numbers in italics represent the lowest values respectively. Plots of annual R^2 can be found in Appendix 7.5.

---

198 See Cook and Weisberg (1983). The results of the Cook-Weisberg test are shown in the respective tables of the regressions.

199 See Whelan and McNamara (2004: 10). The sample is characterised by a large number of pool members (cross section) relative to the total number of observations.

200 See Figure 33 on page 164 for an illustration of the changes in the macroeconomic and capital market environment during the sample period; some other studies run annual estimations to account for such a timely variation, see e.g. Brief and Zarowin (1999); Barth et al. (1998).

Table 9. Value relevance of different accounting variables

Regression models (yearly dummies are not explicitly listed)

$$Y_{it} = b_0 + b_1 \cdot X_{it} + \varepsilon_{it} \qquad \text{(A)}$$

$$Y_{it} = b_0 + b_1 \cdot X_{it} + \varepsilon_{it} \quad \text{only for positive } X_{it} \qquad \text{(B)}$$

$$Y_{it} = b_1 \cdot X_{it} + \varepsilon_{it} \quad \text{only for positive } X_{it} \qquad \text{(C)}$$

where Y is a replacement character for the dependent variables PRICE or EV and X is a replacement character for the independent variables below

Regression results [a]

Variables[b]	Equation (A) R^2	Equation (B) R^2	Equation (C) Uncentred R^2
Dependent Variable: PRICE			
EQUT	**0.378**	0.277	0.663
NI	0.320	**0.402**	**0.725**
FCFE	0.215	0.208	0.631
CFO	0.342	0.274	0.656
Dependent Variable: EV			
SALES	**0.758**	0.750	0.840
EBIT	0.450	0.695	0.810
EBITDA	0.666	**0.839**	**0.900**

[a] Centred R^2 resp. uncentred R^2 is reported for the three models. Uncentred R^2 is not comparable to centred R^2. Results of regressions that use EV as the dependent variable are not comparable to the results of regressions that use PRICE as the dependent variable. Highest values for each model and each dependent variable are indicated in bold letters. Plots of annual R^2 can be found in Appendix 7.5.

[b] Variables are as defined in section 5.2.1 and 5.2.2.

Tables 10 and 11 present the incremental R^2 of several accounting variables given other variables, i.e. the contribution of one variable X_1 to the R^2 of a model that applies both to X_1 and X_2 as independent variables.

Table 10. Incremental R^2 of equity value variables

Incremental R^2 of X_1 given X_2 [a]

Variable X_2 / Variable X_1	EQUT	NI	FCFE	CFO	**Average contribution**
EQUT	-	0.107	0.162	0.048	**0.105**
NI	0.049	-	0.138	0.048	**0.078**
FCFE	0	0.034	-	0.004	**0.012**
CFO	0.012	0.069	0.130	-	**0.070**

[a] Dependent variable: PRICE. Results of regressions that use PRICE as the dependent variable are not comparable to the results of regressions that use EV as the dependent variable.

Table 11. Incremental R^2 of enterprise value variables

Incremental R^2 of X_1 given X_2 [a]

Variable X_2 / Variable X_1	SALES	EBIT	EBITDA	FCFF	TA	**Average contribution**
SALES	-	0.330	0.163	0.192	0	**0.171**
EBIT	0.024	-	0.070	0.066	0.005	**0.040**
EBITDA	0.072	0.285	-	0.141	0.004	**0.126**
FCFF	0.056	0.236	0.097	-	0	**0.097**
TA	0.182	0.493	0.277	0.318	-	**0.317**

[a] Dependent variable: EV. Results of regressions that use EV as the dependent variable are not comparable to the results of regressions that use PRICE as the dependent variable.

As regards Hypothesis 1.1, one can conclude that there is not much difference in the value relevance of *NI* and *CFO*. Indeed, *CFO* has a slightly higher R^2 if the full sample is applied but, if only positive observations are considered, the R^2 of *NI* exceeds the R^2 of *CFO*. Moreover, the uncentred R^2 of model C that suppresses the intercept and the average incremental R^2 is also higher for *NI* than for *CFO*. Furthermore, *NI* has the highest value relevance of all equity models using only positive observations. This clearly emphasizes the dominant role of *NI* as a basis of reference in single-factor equity CCV models (where only positive reference variables are applied). However, because of the full sample dominance of *CFO* it does not seem sufficient to clearly reject the null hypothesis in favour of the alternative hypothesis stated as Hypothesis 1.1. The relatively high value relevance of *CFO* is somewhat surprising in light of the findings of most previous international research. The author believes that this is less an indicator for bad management accounting or weak accounting rules (since German managers cannot be assumed to be more "discretionary" than other managers, and the accounting systems are comparable in all these studies). Instead it is the result of the relatively high number of young companies in the sample, which has been already shown to be a value relevance driver of cash flows (see Black, 1998).

Hypothesis 1.2 states that the value relevance of free cash flow to equity is lower than the value relevance of all the other equity variables. In fact, the tables show that R^2 is lowest for FCFE in all estimation models and that FCFE's average incremental R^2 (=0.012) is also lower than that of other variables. This is consistent with theoretical predictions and clearly indicates that FCFE is not a very suitable variable in CCV. The low relative incremental R^2 of FCFF (=0.097) on the enterprise level supports these findings and emphasizes the inappropriateness of free cash flow as a reference variable in general, even if *EBIT* still has a comparatively lower incremental R^2.

With respect to enterprise value variables, Hypothesis 2 predicts that *SALES* should have lower value relevance than *EBIT* or *EBITDA*. The estimation results, however, cannot support this hypothesis. *EBIT,* but not *SALES* has the lowest value relevance (measured by R^2) in all three regression models. Furthermore, *EBIT* (and also *EBITDA*) exhibits a much lower average incremental R^2 than *SALES*. This is a surprising result. It can partly be justified with the important role of SALES multiples during the new economy boom and with the conclusion that depreciation and amortization do not reflect the true economic value of periodical cost allocation, however, there still remains some mystery about this finding. While *SALES* have also the highest R^2 for the full sample regression, it is ranked second behind EBITDA for both only-positive-variables regressions. This shows that EBITDA – if used as a multiple – is the number one performance-oriented enterprise value reference variable.

As regards the regression analysis thus far, particularly two points merit a comment. First, it is important to note that the results drawn from the regressions that use the enterprise value as the dependent variable are *by no means* comparable to the results that use the stock price as the dependent. This is the case because the enterprise value is calculated using capital market based *and* accounting based figures (note that the market value of debt is proxied by the book value of debt) while stock prices are pure capital market based variables. The consequence is a systematic upward bias of R^2 for the enterprise value regressions. This problem becomes especially clear when looking at a regression model where enterprise value is explained by total assets. In such a model, the book value of debt is part of the dependent and the independent variable. In this context, it can be shown that for very high book values of debt relative to *PRICE* and *EQUT*, R^2 converges to

1.[201] This is also one of the explanations for the relatively high incremental R^2 of *TA* (as reported in Table 11). Thus, only as long as the dependent variable is consistently defined, comparisons are possible. Consequently, results from the regressions with enterprise value as the dependent variable can only be compared to other regression results with enterprise value as the endogenous variable. Moreover, since company valuation usually focuses ultimately on equity investors and due to the methodological problems associated with enterprise value regressions, the regressions with stock price as the dependent variable are more relevant here and allow for a more accurate interpretation of economic reality.

Second, even if correlations of several enterprise value variables are high (e.g. the correlation between *SALES* and *TA* is at 0.9 following the Pearson coefficient and at 0.88 following the Spearman coefficient; see Table 8), a possible multicollinearity does not affect statistical inference when determining the incremental R^2. This is the case since multicollinearity has influence on the standard errors but not the R^2. Thus, no transformation of variables was necessary for these calculations.

Hypothesis 3 *The relationship between the PB ratio and the return on equity is non-linear (i.e. rather convex)*

Some "eyeball econometrics" will set the starting point to this analysis. A graphical look at the two-dimensional relationships between the PB ratio and ROE – presented in Figure 34 – reveals that the plot is consistent with the predicted shape. In order to technically address this hypothesis, the differences in the slope coefficients between the sectors are tested for significance. [202] The highest 3 % of all observations (in terms of ROE) are deleted to reduce the impact of outliers. Table 12 contains the results of these tests.

The coefficient values for the sections are continuously rising from slightly negative (S1) to highly positive (S3). Moreover, the differences between coefficients are significant in all three possible cases. This clearly indicates that the PB-ROE association has a convex shape. Thus, the null hypothesis of a linear relationship can be rejected. This result is consistent with the findings of Burgstahler and Dichev (1997). One important result is that the existence of convexity in the PB-ROE model strongly supports the validity of the two factor model that has been derived in section 4.3.2.

---

201 See Appendix 7.2 for a proof of the convergence of R^2 in simple linear regression models. However, a noticeable upward bias is existent even for low book values of debt. The argument also holds for other models that use the enterprise value as the dependent variable.

202 Consistently with a similar study of Burgstahler and Dichev (1997), the domain of *ROE* is divided into three parts with an equal number of observations.

Fig. 34. Plot of the PB-ROE Association

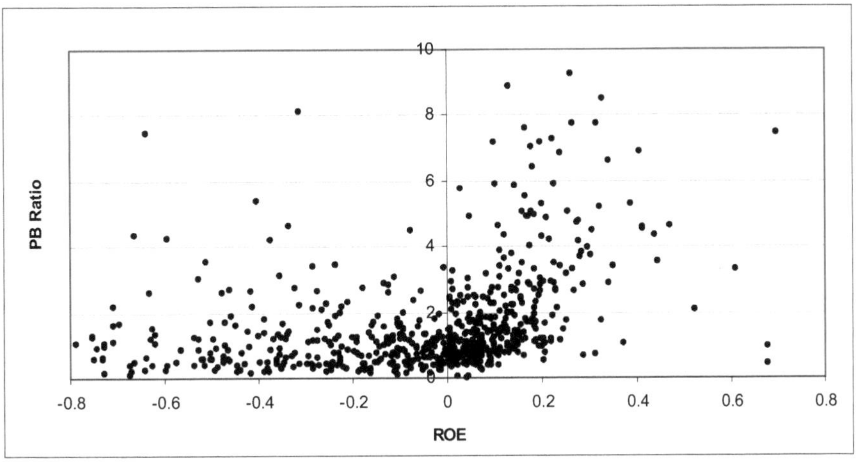

Source: KKMDB, Hoppenstedt.

Table 12. Convexity in the PB-ROE Association

			Coefficients in sections[c]		
Depend. variable[a]	Indep. variable[a]	Cook-Weisberg	S1	S2	S3
PB	ROE	***	-0.24	2.03**	11.12***
			Tests of significant differences in coefficients[b,c] between sections		
			S2-S1	S3-S2	S3-S1
			2.1**	3.7***	4.9***

[a] Variables are as defined in section 5.2.1.

[b] t-statistics are reported; assumption: $\operatorname{cov}\left(b_{SectionA}; b_{SectionB}\right) = 0$.

[c] * (**, ***): significant at the two-tailed 10% (5%, 1%) level; standard errors are heteroscedasticity-corrected (White 1980); t-values are standard normally distributed.

What is noteworthy and surprising, however, is that the coefficient for the low-*ROE* group has a negative sign, which – at the first glance – does not seem to be consistent with financial theory. A possible explanation for this is that very low values are often the result of extraordinary one-time events and therefore lack persistence.[203] Anyway, the unexpected sign of this coefficient should not be stressed too much since it is not significantly different from zero.

---

203 Jan and Ou (1995) and Collins et al. (1999) also found that – when price is regressed on earnings – the coefficient on earnings is reliably negative for loss firms. While the first ones call it a “bewildering phenomenon” the second ones suggest that it is the

Hypothesis 4.1 *The joint value relevance of earnings and book value is higher than the value relevance of each of these two variables alone*

Hypothesis 4.1 is simply tested by comparing the valuation roles of *EQUT* and *NI* in simple linear regressions and multiple linear regressions. The results are presented in Table 13.

Obviously, the adjusted R^2 of regression model C (the model that uses both *NI* and *EQUT* as independent variables) exceeds the R^2 of each of the other two models. This leads to the conclusion that *EQUT* and *NI* have a joint value relevance that is higher than the value relevance of each of these two accounting figures alone. Moreover, the coefficients of *NI* and *EQUT* are highly significant in every model, which provides further support for the validity of the conclusion. It is also important to note that four out of five yearly dummy variables are significantly different from zero. This emphasizes the importance of including time dummy variables into the analysis in general, but it also shows that the sample has often been subject to heavy macro-influences during the sample period.

Hypothesis 4.2 *The value relevance of earnings is reduced and the value relevance of book value is increased for firms that operate in highly competitive industries compared to firms that do not*

Hypothesis 4.3 *The value relevance of earnings is reduced and the value relevance of book value is increased for firms that are asset heavy compared to firms that are not*

Hypothesis 4.4 *The value relevance of earnings is reduced and the value relevance of book value is increased for firms that have a high relative creditworthiness compared to firms that have a poor relative creditworthiness*

These hypotheses state that *EQUT* and *NI* have varying value relevance dependent on the industry structure as well as the creditworthiness and asset heaviness of companies. Tables 14-16 present the results of both the sub-sample comparisons and the full-sample approach using dummy variables. The results are given separately for each of the relevant research questions. Table 17 contains the results if earnings adjusted creditworthiness (ACRW) is used instead of pure creditworthiness (CRW).

As regards Hypothesis 4.2 (Table 14), the results are straightforward. In the competitive industry sub-sample (D1=1) the coefficient on *EQUT* is highly significant while that on *NI* is not. Contrary to that, in the sub-sample that comprises all other industries (D1=0) the coefficient of *NI* is highly significant while *EQUT* is not. Moreover, coefficients of *NI* are higher in the (D1=0) sub-sample than in the (D1=1) sub-sample and coefficients of *EQUT* are lower in the (D1=0) sub-sample than in the (D1=1) sub-sample. These differences are highly significant in both cases: The t-statistics is at -3.58*** in the test of equality of the slope coefficients of *NI* in both sub-sample regressions, and at 2.81*** in the test of equality

---

omission of book value which induces the negative bias. However, in the present analysis book value is not omitted and the sign is still negative.

of the slope coefficients of *EQUT* in both sub-sample regressions (tests are performed under the assumptions: $\text{cov}\left(b_{\text{NI, Equation A}}; b_{\text{NI, Equation B}}\right) = 0$ and $\text{cov}\left(b_{\text{EQUT, Equation A}}; b_{\text{EQUT, Equation B}}\right) = 0$). The incremental R^2 of *NI* given *EQUT* is distinctly higher for the group of companies that are characterised by $D1 = 0$ than for the other sub-sample. The opposite is true for the incremental R^2 of *EQUT* given *NI*: It is much higher for the competitive sub-sample than for the other sub-sample.

Table 13. The general valuation roles of book value and earnings

Regression models (yearly dummies are not explicitly listed)

$$P_{it} = b_0 + b_1 \cdot EQUT_{it} + \varepsilon_{it} \quad \text{(A)}$$
$$P_{it} = b_0 + b_1 \cdot NI_{it} + \varepsilon_{it} \quad \text{(B)}$$
$$P_{it} = b_0 + b_1 \cdot EQUT_{it} + b_2 \cdot NI_{it} + \varepsilon_{it} \quad \text{(C)}$$

Regression results [a,b]

	Equation		
Variables [c]	(A)	(B)	(C)
Cook-Weisberg	***	***	***
Intercept	27.94***	33.08***	28.27***
EQUT	0.56***		0.42***
NI		2.31***	1.36***
DY99	-	-	-
DY00	-14.95***	-14.50***	-13.97***
DY01	-20.05***	-19.03***	-18.65***
DY02	-25.21***	-23.70***	-23.64***
DY03	-19.50***	-20.13***	-18.90***
Adj. R^2	0.378	0.320	0.427
Incremental R^2 (NI given EQUT)			0.049
Incremental R^2 (EQUT given NI)			0.107

[a] Coefficients are reported for the variables; * (**, ***): significant at the two-tailed 10% (5%, 1%) level. Standard errors are heteroscedasticity-corrected (White 1980); t-values are standard normally distributed.
[b] Only significant time dummy variables are reported. Insignificant ones are dropped.
[c] Variables are as defined in section 5.2.1.

Table 14. Impact of industry structure on the relative valuation roles of book value and earnings

Regression models (yearly dummies are not explicitly listed)

$$P_{it} = b_0 + b_1 \cdot EQUT_{it} + b_2 \cdot NI_{it} + \varepsilon_{it} \qquad \text{(A)}$$

sub-sample: competitive industries

$$P_{it} = b_0 + b_1 \cdot EQUT_{it} + b_2 \cdot NI_{it} + \varepsilon_{it} \qquad \text{(B)}$$

sub-sample: other industries

$$P_{it} = b_0 + b_1 \cdot D1_i + b_2 \cdot EQUT_{it} + b_3 \cdot D1_i \cdot EQUT_{it} + + b_4 \cdot NI_{it} + b_5 \cdot D1_i \cdot NI_{it} + \varepsilon_{it} \qquad \text{(C)}$$

Regression results [a,b]

	Equation		
Variables[c]	(A)	(B)	(C)
Cook-Weisberg	***	***	***
Intercept	13.68***	31.03***	32.10***
D1			-9.51***
EQUT	0.68***	0.17	0.17
NI	0.64	2.92***	2.92***
D1 EQUT			0.51***
expected sign			positive
D1 NI			-2.26***
expected sign			negative
DY99	11.35***	-	-
DY00	-3.67**	-9.36**	-11.75***
DY01	-9.47***	-14.10***	-16.95***
DY02	-11.97***	-22.77***	-22.25***
DY03	-8.35***	-16.59***	-17.41***
Adj. R^2	0.473	0.414	0.469
Incremental R^2 (NI given EQUT)	0.019	0.111	
Incremental R^2 (EQUT given NI)	0.212	0.017	

Test EQUT: [d]

H_0: $b_{1,\text{Equation Xa}} = b_{1,\text{Equation Xb}}$ t-statistic : 2.81***

Test NI: [d]

H_0: $b_{2,\text{Equation Xa}} = b_{2,\text{Equation Xb}}$ t-statistic : -3.58***

[a] Coefficients are reported for the variables; * (**, ***): significant at the two-tailed 10% (5%, 1%) level. Standard errors are heteroscedasticity-corrected (White 1980); t-values are standard normally distributed.

[b] Only significant dummy variables are reported. Insignificant ones are dropped.

[c] Variables are as defined in section 5.2.1 and 5.2.3. $D1_i$=1 if industry is competitive, 0 otherwise.

[d] Assumption: $\text{cov}\left(b_{\text{Equation A}}; b_{\text{Equation B}}\right) = 0$

The full sample regression (model C in Table 14) shows that the coefficient on the intercept dummy is significant suggesting that industry structure is value relevant in its own right. The coefficient of the dummy variable multiplied by *NI* is significantly negative (-2.26***) and the dummy multiplied by *EQUT* is significantly positive (0.51***), indicating that the market's reaction to *NI* is reduced and to *EQUT* is increased for companies in competitive industries relative to other industries. Again, most of the yearly dummies are highly significant in all three regressions indicating that the sample is affected by timely varying macro-influences.

These results indicate that the value relevance of earnings is lower for competitive industries as compared to other industries. Likewise, the value relevance of the book value of equity is higher in competitive industries as compared to other industries. This clearly supports Hypothesis 4.2. The message in terms of CCV is the following: Book value multiples are expected to be the better tool compared to earnings multiples if the target company is operating in a competitive industry (and vice versa if the company is not operating in a competitive industry).

As regards Hypothesis 4.3 (Table 15), the results are no longer quite so straightforward. In fact, in the asset heavy companies sub-sample (D2=1), the coefficient of *NI* is lower and the coefficient of *EQUT* is higher than in the (D2=0) sub-sample. However, the differences are only significant for the *EQUT* coefficients: The t-statistic is at 2.08** in the test of equality of the slope coefficients of *EQUT* in both sub-sample regressions, and at -0.42 (not significant at the 10% level) in the test of equality of the slope coefficients of *NI* in both sub-sample regressions (tests are performed under the assumptions:

$\mathrm{cov}\left(b_{\mathrm{NI,Equation\,A}};b_{\mathrm{NI,Equation\,B}}\right)=0$ and

$\mathrm{cov}\left(b_{\mathrm{EQUT,Equation\,A}};b_{\mathrm{EQUT,Equation\,B}}\right)=0$). The incremental R^2 of *NI* given *EQUT* is slightly higher for the group of companies that are characterised by D2=0 than for the other sub-sample – however both at a very low level. Contrary to that, the incremental R^2 of *EQUT* given *NI* is much higher for the asset heavy sub-sample than for the other sub-sample.

The full sample regression (model C in Table 15) shows that the coefficient on the intercept dummy is significant suggesting that asset heaviness is value relevant in its own right. The coefficient of the dummy variable multiplied by *NI* is not significant but the dummy multiplied by *EQUT* is significantly positive (0.32***), indicating that market's reaction to *EQUT* is increased for asset heavy companies relative to other companies. As in the industry structure analysis, most of the yearly dummies are highly significant in all three regressions indicating that the sample is affected by varying macro-influences over time.

These results indicate that the role of book value in CCV is more important for asset heavy companies than for other companies. However, the increase in the value relevance of the book value of equity is not offset by a decrease in earnings value relevance. Thus, it does not matter for the valuation role of earnings whether a company is asset heavy or not. This may reflect the markets' fixation on earnings – independently of the nature of the company – but also the better "valuability" of asset heavy companies in general.

Table 15. Impact of corporate asset heaviness on the relative valuation roles of book value and earnings

Regression models (yearly dummies are not explicitly listed)

$$P_{it} = b_0 + b_1 \cdot EQUT_{it} + b_2 \cdot NI_{it} + \varepsilon_{it} \quad \text{(A)}$$

sub-sample: asset heavy companies

$$P_{it} = b_0 + b_1 \cdot EQUT_{it} + b_2 \cdot NI_{it} + \varepsilon_{it} \quad \text{(B)}$$

sub-sample: other companies

$$P_{it} = b_0 + b_1 \cdot D2_{it} + b_2 \cdot EQUT_{it} + b_3 \cdot D2_{it} \cdot EQUT_{it} + + b_4 \cdot NI_{it} + b_5 \cdot D2_{it} \cdot NI_{it} + \varepsilon_{it} \quad \text{(C)}$$

Regression results [a,b]

	Equation		
Variables[c]	(A)	(B)	(C)
Cook-Weisberg	***	***	***
Intercept	21.48***	33.08***	29.94***
D2			-5.40***
EQUT	0.66***	0.29*	0.30**
NI	1.06***	1.44*	1.28***
D2 EQUT			0.32**
Expected sign			positive
D2 NI			-
expected sign			negative
DY99	-	-	-
DY00	-11.04***	-16.12***	-13.44***
DY01	-15.60***	-21.45***	-18.31***
DY02	-20.19***	-26.71***	-23.35***
DY03	-15.42***	-22.00***	-18.63***
Adj. R^2	0.598	0.326	0.448
Incremental R^2 (NI given EQUT)	0.027	0.060	
Incremental R^2 (EQUT given NI)	0.218	0.051	

Test EQUT: [d]

H_0: $b_{1,\text{Equation Xa}} = b_{1,\text{Equation Xb}}$ t-statistic : 2.08**

Test NI: [d]

H_0: $b_{2,\text{Equation Xa}} = b_{2,\text{Equation Xb}}$ t-statistic : -0.42

[a] Coefficients are reported for the variables; * (**, ***): significant at the two-tailed 10% (5%, 1%) level. Standard errors are heteroscedasticity-corrected (White, 1980); t-values are standard normally distributed.
[b] Only significant dummy variables are reported. Insignificant ones are dropped.
[c] Variables are as defined in section 5.2.1 and 5.2.3. $D2_{it}=1$ if company is asset heavy, 0 otherwise.

[d] Assumption: $cov\left(b_{Equation\,A}; b_{Equation\,B}\right) = 0$

The results with respect to Hypothesis 4.4 are presented in Tables 16 and 17. First, the impact of creditworthiness on the relative value relevance of earnings and book value is examined (Table 16). The regression results are not very meaningful. The sub-sample comparisons reveal that the coefficient on book value is higher (significant at the 10% level assuming $cov\left(b_{EQUT,Equation\,A}; b_{EQUT,Equation\,B}\right) = 0$) for companies with low creditworthiness relative to companies with high creditworthiness. Moreover, the incremental R^2 of *EQUT* given *NI* is higher for less creditworthy companies than for highly creditworthy companies.[204] However, the full sample regression does not confirm these findings. If the two sub-samples are commonly estimated, the coefficient of the dummy multiplied by *EQUT* is not significant. With respect to the valuation role of *NI*, no significant differences have been found for the two sub-samples – neither in the sub-sample comparison nor in the full sample regression. Nevertheless, the coefficient on the intercept dummy is significant indicating that creditworthiness is at least value relevant in its own right. Many time dummies are significant. The weak indications for a dominating valuation role of book value for companies with low creditworthiness are probably a sign of book value's nature as a liquidation value proxy. However, the evidence is too weak to draw valid conclusions and to reject the null hypothesis (H_0: no influence of creditworthiness on the relative valuation roles of earnings and book value). This result is an extensive indicator for the offsetting nature of the two conflicting theories about the impact of creditworthiness on the valuation roles of accounting variables.

Things change dramatically, however, if the dummy is no longer defined over pure creditworthiness but over earnings adjusted creditworthiness (Table 17). The coefficient on *NI* is lower (significant at the 1% level assuming $cov\left(b_{NI,Equation\,A}; b_{NI,Equation\,B}\right) = 0$) for companies with high adjusted creditworthiness (D4=1) relative to companies with low adjusted creditworthiness (D4=0). Moreover, the incremental R^2 of *NI* given *EQUT* is more than twice as high for (D4=0) companies than it is for the other companies. The full sample regression (model C in Table 17) shows that the coefficient of the dummy variable multiplied by *NI* is significantly negative. Furthermore, the coefficient on the intercept dummy is significant suggesting that earnings adjusted creditworthiness is value relevant in its own right. With respect to the relative value relevance of

---

204 This is largely consistent with the findings of Barth et al. (1998).

Table 16. Impact of creditworthiness on the relative valuation roles of book value and earnings

Regression models (yearly dummies are not explicitly listed)

$$P_{it} = b_0 + b_1 \cdot EQUT_{it} + b_2 \cdot NI_{it} + \varepsilon_{it} \quad \text{(A)}$$

sub-sample: high creditworthiness

$$P_{it} = b_0 + b_1 \cdot EQUT_{it} + b_2 \cdot NI_{it} + \varepsilon_{it} \quad \text{(B)}$$

sub-sample: low creditworthiness

$$P_{it} = b_0 + b_1 \cdot D3_{it} + b_2 \cdot EQUT_{it} + b_3 \cdot D3_{it} \cdot EQUT_{it} + + b_4 \cdot NI_{it} + b_5 \cdot D3_{it} \cdot NI_{it} + \varepsilon_{it} \quad \text{(C)}$$

Regression results [a,b]

	Equation		
Variables[c]	(A)	(B)	(C)
Cook-Weisberg	***	***	***
Intercept	25.08***	26.57***	25.57***
D3			7.79***
EQUT	0.31*	0.64***	0.59***
NI	1.93**	1.19**	1.23**
D3 EQUT			-
Expected sign			no expectation
D3 NI			-
Expected sign			no expectation
DY99	-	-	-
DY00	-6.60*	-14.06***	-13.21***
DY01	-	-19.62***	-17.99***
DY02	-11.29**	-23.99***	-22.79***
DY03	-	-20.32***	-18.25***
Adj. R^2	0.231	0.454	0.477
Incremental R^2 (NI given EQUT)	0.043	0.042	
Incremental R^2 (EQUT given NI)	0.037	0.1725	
Test EQUT: [d] H_0: $b_{1,\text{Equation Xa}} = b_{1,\text{Equation Xb}}$	t-statistic : -1.81*		
Test NI: [d] H_0: $b_{2,\text{Equation Xa}} = b_{2,\text{Equation Xb}}$	t-statistic : 0.66		

[a] Coefficients are reported for the variables; * (**, ***): significant at the two-tailed 10% (5%, 1%) level. Standard errors are heteroscedasticity-corrected (White, 1980); t-values are standard normally distributed.

[b] Only significant dummy variables are reported. Insignificant ones are dropped.

[c] Variables are as defined in section 5.2.1 and 5.2.3. $D3_{it}$=1 if company has a high creditworthiness, 0 otherwise.

[d] Assumption: $\text{cov}\left(b_{\text{Equation A}}; b_{\text{Equation B}}\right) = 0$

Table 17. Impact of earnings adjusted creditworthiness on the relative valuation roles of book value and earnings

Regression models (yearly dummies are not explicitly listed)

$$P_{it} = b_0 + b_1 \cdot EQUT_{it} + b_2 \cdot NI_{it} + \varepsilon_{it} \quad \text{(A)}$$

sub-sample: high adj. creditworthiness

$$P_{it} = b_0 + b_1 \cdot EQUT_{it} + b_2 \cdot NI_{it} + \varepsilon_{it} \quad \text{(B)}$$

sub-sample: low adj. creditworthiness

$$P_{it} = b_0 + b_1 \cdot D4_{it} + b_2 \cdot EQUT_{it} + b_3 \cdot D4_{it} \cdot EQUT_{it} + b_4 \cdot NI_{it} + b_5 \cdot D4_{it} \cdot NI_{it} + \varepsilon_{it} \quad \text{(C)}$$

Regression results [a,b]

	Equation		
Variables[c]	(A)	(B)	(C)
Cook-Weisberg	***	***	***
Intercept	22.00***	28.91***	25.59***
D4			7.85***
EQUT	0.56***	0.53***	0.54***
NI	0.68	2.55***	2.53***
D4 EQUT			-
Expected sign			positive
D4 NI			-1.80***
Expected sign			negative
DY99	-	-	-
DY00	-6.35*	-15.07***	-12.47***
DY01	-	-20.29***	-16.75***
DY02	-13.09***	-26.39***	-23.44***
DY03	-	-21.68***	-17.73***
Adj. R^2	0.339	0.526	0.498
Incremental R^2 (NI given EQUT)	0.029	0.084	
Incremental R^2 (EQUT given NI)	0.113	0.100	
Test EQUT:[d] H_0: $b_{1,\text{Equation Xa}} = b_{1,\text{Equation Xb}}$	t-statistic : 0.29		
Test NI:[d] H_0: $b_{2,\text{Equation Xa}} = b_{2,\text{Equation Xb}}$	t-statistic : -2.99***		

[a] Coefficients are reported for the variables; * (**, ***): significant at the two-tailed 10% (5%, 1%) level. Standard errors are heteroscedasticity-corrected (White, 1980); t-values are standard normally distributed.

[b] Only significant dummy variables are reported. Insignificant ones are dropped.

[c] Variables are as defined in section 5.2.1 and 5.2.3. $D4_{it}=1$ if company has a high earnings adjusted creditworthiness, 0 otherwise.

[d] Assumption: $\text{cov}\left(b_{\text{Equation A}}; b_{\text{Equation B}}\right) = 0$

EQUT, no significant differences have been found for the two states of adjusted creditworthiness – neither in the sub-sample comparison nor in the full sample regression. Again, most of the yearly dummies are highly significant especially in the low adjusted creditworthiness sub-sample regression and the full sample regression.

These results indicate that creditworthiness – if adjusted for the value of earnings – has a negative impact on the value relevance of earnings but no impact on book value. If used as an unadjusted measure, however, creditworthiness has a slightly negative impact on book value but no impact on earnings. The consequences for CCV are straightforward: There are some weak indications that the use of book value multiples is more promising if the target company is of low creditworthiness than if it is of high creditworthiness. If, however, two companies have the same level of (normalised) earnings, then the valuation accuracy of earnings multiples is probably higher for the low creditworthy company than for the high creditworthy company.

Robustness checks of Hypotheses 4.2, 4.3 and 4.4

To assess the robustness of the results with respect to Hypotheses 4.2, 4.3 and 4.4, a "general" model that contains all variables used in the previous analyses is established. The results that are presented in Table 18 (model A) merit some comment. First, industry structure still has a significant negative influence on earnings relevance but the influence on book value is no longer significant. Industry structure is still value relevant in its own right. Second, asset heaviness is still of relevance in its own right, but no longer has any significant influence on the value relevance of book value (remember that the separate analysis about asset heaviness did not show any impact on the value relevance of earnings already).[205] Third, unadjusted creditworthiness is still value relevant itself (remember that the separate analysis about creditworthiness did not show any impact on the value relevance of book value and earnings already). Fourth, earnings adjusted creditworthiness is no longer of relevance in its own right but it still has a highly significant influence on the value relevance of earnings (remember that the separate analysis about adjusted creditworthiness did not show any impact on the value relevance of book value already).[206] As a result it turns out that only industry structure's and adjusted creditworthiness' influence on earnings value relevance remain significant (the signs of the remaining significant coefficients are still as expected). It is important to note that this result highlights the strong impact of these two dummy variables but contemporaneously does not call into question the previous analyses about asset heaviness and unadjusted creditworthiness. In fact, the general model only provides evidence as to which influences dominate other influences, and

205 A possible explanation for this is that companies with high capital intensity typically operate in low competition industries, see White et al. (1997: 189-190).

206 It is concluded that multicollinearity is not a problem here since the F-statistic (which indicates overall significance of the model) corresponds with the t-statistics on coefficients (which indicate significance of the remaining variables).

which influences have to be considered if more than one of the above named circumstances applies at the same time.

Recent research has identified several variables that also affect the value relevance of book value and earnings (see e.g. Barth et al., 1998; Collins et al., 1999; Brief and Zarowin, 1999; Whelan and McNamara, 2004). In order to further examine the robustness of the general model, these factors are controlled for through inclusion of additional independent variables. More precisely, the following control variables are part of an extended general model (model B in Table 18):

(1) *GDP* (Germany) for the years 1998-2003, (2) *GDPEXPECT* (Germany) for the years 1998-2003, (3) *BETA*, (4) $EQUT\,/\,TA$ (as a measure of leverage), (5) *ROE*, (6) *NINEG* (to control for possible non-linearities in the relationship between price and book value or earnings), (7) *TA* and finally (8) *DIV*.[207]

The control for these variables is necessary to draw valid results since the *omitted variables problem* (i.e. relevant information is omitted from the model) is more serious than the *overspecification problem* (i.e. assigning variables to hypotheses that are not immediately relevant). Following a general-to-specific procedure, all insignificant variables are successively dropped so that only significant variables are reported (exception: $\mathrm{D1} \cdot \mathrm{NI}$ is reported although no longer significant). What is noteworthy here is that several intercept and time dummies become insignificant. This is offset by the significant control variables *DIV*, *NINEG* and *TA*.

The signs are largely as expected. Only the negative sign of the *TA*-coefficients is somewhat surprising since it suggests stock prices will increase as a firm's amount of assets decreases. The influence of earnings adjusted creditworthiness on earnings value relevance is still highly significant but the influence of industry structure on earnings relevance is not. However, it is worth noting that the coefficient on $\mathrm{D1} \cdot \mathrm{NI}$ is still significantly negative (one tailed) on a 10% level of significance.[208]

To summarise: The inclusion of the control variables in the regression had only a weak effect on the relevant coefficients and hypothesis tests, which is a sign of robustness. Although industry structure's influence on earnings relevance becomes slightly insignificant in two-tailed tests, it still is significantly negative in a one-tailed 10% level.

---

207 Note: GDP and GDPEXPECT do not vary in the cross-section.

208 Again, it is concluded that multicollinearity is not a problem here since the F-statistic (which indicates overall significance of the model) corresponds with the t-statistics on coefficients (which indicate significance for the remaining variables).

Table 18. Complete regression model and robustness check on the relative valuation roles of book value and earnings

Regression models (yearly dummies are not explicitly listed)

$$P_{it} = b_0 + b_1 \cdot D1_i + b_2 \cdot D2_{it} + b_3 \cdot D3_{it} + b_4 \cdot D4_{it} + +b_5 \cdot EQUT_{it} + b_6 \cdot NI_{it} + +b_7 \cdot D1_i \cdot NI_{it} + b_8 \cdot D2_{it} \cdot NI_{it} + b_9 \cdot D3_{it} \cdot NI_{it} + b_{10} \cdot D4_{it} \cdot NI_{it} + +b_{11} \cdot D1_i \cdot EQUT_{it} + b_{12} \cdot D2_{it} \cdot EQUT_{it} + b_{13} \cdot D3_{it} \cdot EQUT_{it} + b_{14} \cdot D4_{it} \cdot EQUT_{it} + \varepsilon_{it} \quad \text{(A)}$$

The same model but additionally including several control variables (B)

Regression results [a,b]

Variables[c]	Equation (A)	Equation (B)
Intercept	28.26***	16.69***
D1	-3.39***	-
D2	-2.98***	-
D3	9.06***	7.16***
D4	-	-
EQUT	0.53***	0.73***
NI	2.73***	1.80***
D1 EQUT	-	-
D1 NI	-1.08*	-0.68 (p-value: 0.196)
D2 EQUT	-	-
D2 NI	-	-
D3 EQUT	-	-
D3 NI	-	-
D4 EQUT	-	-
D4 NI	-1.44***	-1.27***
DIV		9.08***
NINEG		-5.16***
TA		-0.13***
DY99	-	-
DY00	-10.62***	-
DY01	-15.87***	-
DY02	-22.52***	-9.75***
DY03	-17.32***	-4.87***
Adj. R^2	0.511	0.516

[a] Coefficients are reported for the variables; * (**, ***): significant at the two-tailed 10% (5%, 1%) level. Standard errors are heteroscedasticity-corrected (White, 1980); t-values are standard normally distributed.

[b] Only significant dummy variables are reported. Insignificant ones are dropped.

[c] Variables are as defined in section 5.2.1 and 5.2.3.

5.2.4.3 Results from the Matching Estimator Approach

Hypothesis 5.1 *A company has a lower relative value if it operates in highly competitive industries than if it operates in non-competitive industries – assuming accounting figures are the same in both cases*

Hypothesis 5.2 *A company has a lower relative value if it has a low degree of creditworthiness than if it has a high degree of creditworthiness – assuming accounting figures are the same in both cases*

Table 19 reports the results from the probit estimations.[209] The middle column shows the coefficients of the regression of D1 on the covariates *NI*, *EQUT*, *EQUTRA*, and *SALES* and the rightmost column shows the coefficients of the regression of D3 on these covariates. The basic results are here, that, as expected, firms with lower earnings and book value exhibit a significantly higher probability to be in a competitive industry, and firms with higher earnings and book value exhibit a significantly higher probability to have a good credit rating.

Table 19. Probit estimations on industry structure (D1) and creditworthiness (D3)

	Dependent variable[a,b]	
Variables[b]	D1	D3
Intercept	-0.303***	-0.921***
EQUT	-0.021***	0.022***
NI	-0.070***	0.164***
SALES	0.003**	-0.003***
EQUTRA	1.130***	-0.928***
Log Likelihood	-520.398	-244.484
Aldrich Nelson Pseudo R^2	0.149	0.247

[a] Coefficients are reported for the variables; * (**, ***): significant at the two-tailed 10% (5%, 1%) level.
[b] Variables are as defined in sections 5.2.1 and 5.2.3.

Before discussing the matching results, it is important to go back to one of the basic assumptions of the matching estimator approach: the Conditional Independence Assumption (CIA) (see Rubin, 1977; Gerfin and Lechner, 2000: 18-19). The CIA states in the present context that firms from the treatment group and the control group – with identical realisations of *G* – differ in *PRICE* only through the implications of their treatment. This validity of this assumption is formally

209 It is assumed here that the conditional probabilities follow a normal cumulative distribution.

untestable (see Almus and Czarnitzki, 2001: 11). However, it seems to be reasonable to assume that in the case of Hypothesis 5.1 industry classification (with all its consequences on the covariates *NI*, *EQUT*, *EQUTRA* and *SALES*), and in the case of Hypothesis 5.2 creditworthiness (with all its consequences for the covariates *NI*, *EQUT*, *EQUTRA* and *SALES*) are the only stock-price-relevant differences between both groups. To put it differently: It can be assumed that in both cases the CIA seems to hold.

As has been outlined already, the matching technique is the *Epanechnikow kernel*, i.e. all members of the non-treatment group are used in order to build a match for each member of the treatment group. The contribution of each control group member to that match is a function of the distance of its *propensity score* from the propensity score of the respective treatment group member. Propensity scores are drawn from the probit models (Table 19).

Due to the *common support* restriction, the matching procedure could not assign a match to all treatment group members. In the matching procedure with D1 as the dependent variable, 1 company could not be matched, and in the procedure with D3 as the dependent variable, 3 companies could not be matched (Table 20).

Table 20. Matching success under common support

	Off support	On support	Total
Treatment: Industry structure (D1)			
Untreated	0	372	**372**
Treated	1	437	**438**
Total	**1**	**809**	**810**
Treatment: Creditworthiness (D3)			
Untreated	0	596	**596**
Treated	3	98	**101**
Total	**3**	**694**	**697**

Finally, the matching results are presented in Table 21. The columns that are labelled "Mean Treated" and "Mean Control" show the averages of the outcome variable *PRICE* for the treated (companies for which D1=1 resp. D3=1 holds) and the control group (companies for which D1=0 resp. D3=0 holds). The column labelled "Differences of Sample Means" contains the average difference in the outcome variable *PRICE* between these two groups for the unmatched sample. This result is obtained by a simple mean comparison. The rightmost column contains the average difference in *PRICE* between both groups after the matching. This difference is called the average treatment effect to the treated (ATT), and is the central result of the whole matching procedure.

Table 21. Matching results (Epanechnikow Matching)

	Mean		Difference of sample means	ATT[a]
	Treated	Control		
Treatment: Industry structure (D1); Outcome variable: PRICE				
Unmatched sample	10.72	20.04	-9.32***	
Matched sample	10.74	14.86		-4.12***
Treatment: Creditworthiness (D3); Outcome variable: PRICE				
Unmatched sample	31.81	13.29	18.52***	
Matched sample	31.55	23.23		8.32***

[a] ATT: average treatment effect on the treated; coefficients are reported; * (**, ***): significant at the two-tailed 10% (5%, 1%) level; t-values are standard normally distributed.

The interpretation of matching results is straightforward: A simple mean comparison shows that companies in competitive industries and companies with a low rating have lower relative prices than other companies. After correcting for the selection bias (i.e. for different values in the accounting variables), these differences are still significant. This means that companies in competitive industries have lower relative prices than companies in other industries (assuming both have the same accounting figures). As a consequence, a company trades at lower multiples if it operates in a competitive industry than if it operates in other industries (assuming it has the same value of accounting figures in both cases).[210] Likewise, stock prices are lower for low-rated companies relative to companies with a high credit rating after controlling for the selection bias. This means that – given a set of accounting variables with identical values for all companies – creditworthy companies have higher prices than others, i.e. they trade at higher multiples. These results strongly support Hypotheses 5.1 and 5.2.

Robustness check

To assess the robustness of the matching results, a mean comparison of the covariates is performed after the matching (Table 22). It is desirable that the covariates do not exhibit significant differences between both groups, since differences would indicate that the matching was not successful.

---

210 The price relative to a given set of accounting variables with identical value is nothing else than a multiple.

Table 22. Mean comparison of the covariates after Epanechnikow Matching

Variables[b]	Mean		t-statistic[a] (test of significant differences)
	Treated	Control	
Treatment: Industry structure (D1); Matched sample			
EQUT	7.28	7.88	-0.87
NI	-0.37	-0.22	-1.31
SALES	22.71	23.60	-0.25
EQUTRA	0.51	0.52	-0.70
Treatment: Creditworthiness (D3); Matched sample			
EQUT	18.95	16.60	2.99***
NI	1.80	1.45	3.71***
SALES	65.69	55.00	3.41***
EQUTRA	0.37	0.34	1.89*

[a] * (**, ***): significant at the two-tailed 10% (5%, 1%) level; t-values are standard normally distributed.
[b] Variables are as defined in section 5.2.1.

The table shows that the covariates are not significantly different between both groups in the case where D1 (industry structure) serves as the dependent variable. This is a sign for high robustness of the previous results. However, for the analysis where D3 (creditworthiness) serves as the dependent variable, all covariates are significantly different for both groups even after the matching. This means that the matching process was not very successful, i.e. the two groups are still different with respect to the covariates. As a consequence, a different matching technique is applied for the D3-analysis in order to ensure some minimum level of homogeneity within the matches. This new matching technique is called *Mahalanobis Matching*. The difference between the Mahalanobis matching and the Epanechnikow matching is that the Mahalanobis matching is a one-to-one matching, i.e. each observation of the high-creditworthiness group is matched with one observation of the control group according to the distance of propensity scores (nearest neighbour matching).[211] In addition to the propensity scores, the matches are restricted to show similarity with regard to the covariates *NI*, *EQUT*, *EQUTRA,* and *SALES*. The distance measure used to condition on the four covariates is Maha-

211 As has been mentioned before, the observation distribution for both groups principally allows a nearest neighbour matching here.

lanobis distance.[212] Table 23 reports the matching result and Table 24 shows the new mean comparisons of the covariates.

Table 23. Matching results (Mahalanobis Distance)

	Mean		Difference of sample means	ATT[a]
	Treated	Control		
Treatment: Creditworthiness (D3); Outcome variable: PRICE				
Unmatched sample	31.81	13.29	18.52***	
Matched sample	31.55	25.25		6.30**

[a] ATT: average treatment effect on the treated; coefficients are reported; * (**, ***): significant at the two-tailed 10% (5%, 1%) level; t-values are standard normally distributed.

Table 24. Mean comparison of the covariates after Mahalanobis Matching

Variables[b]	Mean		t-statistic[a] (test of significant differences)
	Treated	Control	
Treatment: Creditworthiness (D3); Matched sample			
EQUT	18.95	18.04	1.04
NI	1.80	1.64	1.52
SALES	65.69	58.98	1.58
EQUTRA	0.37	0.38	-0.52

[a] * (**, ***): significant at the two-tailed 10% (5%, 1%) level; t-values are standard normally distributed.
[b] Variables are as defined in section 5.2.1.

The results from the Mahalanobis Matching and the mean comparison of the covariates speak a clear language: The difference in prices is still significant between both groups while the covariates do not show any significant differences anymore. Obviously, the results of the Mahalanobis matching confirm the results from the Epanechnikow matching in regards to the analyses in which D3 serves as the dependent variable. This finding strongly supports Hypothesis 5.2 and, thus, clearly indicates that highly creditworthy companies trade at higher multiples than low rated companies.

212 For more details about this approach, see Cochrane and Rubin (1973), Rosenbaum and Rubin (1985). The common support restriction does not change under this approach. Thus, the number of companies that are off-support remains at 3. The matching protocol is very similar to the protocol that is presented in figure 32.

5.3 Pricing Accuracy

5.3.1 Previous Empirical Results

In the majority of cases the existing empirical literature about pricing accuracy relates to one of the two main fields of CCV research: the selection of the peer group companies and the choice of an accurate valuation model. The most important studies are briefly described below.

The study of Alford (1992) deals with the selection of the peer group companies when using the PE ratio. This study examines criteria for the US stock market that are important for the composition of the comparable companies group. Alford finds that industry classification is a very efficient criterion. However, the same risk class and a similar earnings growth – if used together – also lead to accurate valuation results although neither variable performs well by itself. One key result of this study is that leverage is not relevant when selecting the peer group companies.

Bhojraj and Lee (2002) also concentrate on the composition of the peer group. Based on cross-sectional variations of variables nominated by valuation theory, they develop a "warranted" EV/SALES and PB ratio for each company. Variables used in this regressional approach include profitability, certain growth measures, and the cost of capital. Suitable peer group companies are defined as those having the warranted multiple closest to that of the target company. The authors conclude that this method is more objective and results in a higher pricing accuracy as compared to other peer group selection approaches.

Dittmann and Weiner (2005) analyse the pricing accuracy of the EV/EBIT ratio for different peer group selection methods. Based on a broad international sample it turns out that profitability is a highly important comparability criterion. Composing the set of comparable companies based on return on assets yields more accurate valuation results than using industry classification or total assets as the comparability basis.

Kaplan and Ruback (1995) examine the performance of the EV/EBITDA ratio for the US market. They compare the valuation accuracy of this ratio with the results of a DCF valuation. Their benchmark is the realised price in highly leveraged transactions. They find that the DCF valuation performs slightly better than the CCV model. However, the CCV results add explanatory power to the DCF approach. They conclude that DCF and valuation using multiples are complementary valuation models and, thus, both types of approaches should be used in practical valuation settings.

The focus of the study of Kim and Ritter (1999) is on the valuation of companies in IPOs in the USA. They compare different single-factor CCV models using a peer group consisting of all companies from the same industry sector as the target company. They find that common trailing CCV models have only modest predictive power if not adjusted for differences in e.g. sales growth or leverage. They conclude that this is the case because the ratios vary widely, especially for young firms. However, valuation results based on forward CCV models are very accurate. Additionally, the valuation accuracy increases with the age of the company.

Baker and Ruback (1999) find that the valuation power of different bases of reference varies across industries in the USA. They suggest that this result arises from the existence of different value drivers for each industry. Furthermore, they conclude that using the harmonic mean rather than the median or the arithmetic mean is more appropriate to estimate the relevant multiple.

Liu, Nissim and Thomas (2002) also focus on the valuation performance of different single-factor CCV models on the US stock market. They find that multiples based on forward earnings generally outperform multiples based on historical earnings. Moreover, earnings-based models have more valuation power than cash flow-based models or asset-based models. Using sales as the basis of reference delivers the least accurate valuation results. They also examine the performance of combined models such as the "book value/return on capital employed" ratio and cannot find any improvement in valuation accuracy as compared to the single-factor models.

Richter and Herrmann (see Richter and Herrmann, 2002; Herrmann, 2002) investigate both, the choice of appropriate bases of reference in single-factor models and the selection of relevant comparable companies. On the basis of an international sample they find that using earnings as a basis of reference leads to more accurate valuation results than using book value, invested capital or sales. Furthermore, their study reveals that industry membership based on SIC industry codes does not seem to be the most suitable criterion for selecting peer group companies. It is shown that a selection based on earnings growth expectations leads to significantly more accurate valuation results.

Choudhary (2004) examines the influence of the accounting quality on the valuation model choice for the US market. It turns out that companies with high earnings' quality can be more accurately priced with the PE or the PB ratio than companies with low earnings quality. However, companies with high cash flow persistence can be more accurately valued with CCV models that use cash based reference variables than companies with low cash flow persistence.

In addition to these studies that mainly concern single-factor CCV, there are few examinations that deal with multi-factor CCV. All of these analyses focus on the joint use of substance-oriented and performance-oriented multiples. Cheng and McNamara (2000) evaluate the valuation accuracy of the PE and the PB ratio and an equally-weighted PE/PB-ratio. The major result is that the combined ratio requires the fewest criteria to select comparable companies and outperforms both single-factor ratios.

Beatty, Riffe and Thompson (1999) investigate the prediction power of different valuation models to perform tax court valuations of private firms in the USA. Among several other ratios they apply a model in their empirical study that regressionally determines the weights of book value of equity and earnings. They find that this model is preferable among all other combined ratios like the equally-weighted median PE/PB ratio or the equally-weighted average PE/PB ratio. Furthermore they show that in general common single factor models underperform the combined models.

5.3.2 Variable Definition and Methodology

The research design of the pricing accuracy study is straightforward and largely corresponds to the design of previous studies. The analysis proceeds in two steps. In the first step, stock price predictions are calculated for each company in each year of the sample period. This is done by applying several single-factor and multi-factor models with different aggregation methods. One of the multi-factor models applied is the two-factor model that was derived in section 4.3.2. In the second step, the pricing error (i.e. the relative deviation of the predicted stock price from the actual stock price) for each firm, year and model is computed. Based on a comparison of the aggregated pricing errors, an assessment of the pricing accuracy of each model is possible. As a matter of course, high pricing errors are a sign of low pricing accuracy and vice versa.

The focus of this study is on models that use book value of equity and/or earnings as reference variables. Thus, the following models are compared to the derived two-factor model: the PE ratio, the PB ratio and an equally-weighted combination model of the PE and PB ratio, each with different aggregation methods. These aggregation methods are the arithmetic mean, the harmonic mean and the median. For technical reasons, these nine models can only be applied to companies – both the target company and the comparable companies – with a positive basis of reference (i.e. positive net income in the case of the PE ratio and positive book value of equity in the case of the PB ratio).[213] Comparable companies are selected – largely according to the practice-oriented selection approach – based on industry classification.[214] An out-of-sample prediction is performed in order to determine the stock price estimates, i.e. the predicted stock price of the target company is calculated based on the assumption that the actual stock price of the target is unknown. This is consistent with the proceeding in valuation reality. Following the recent literature (see Kaplan and Ruback, 1995; Liu et al., 2002; Choudhary, 2005), each firm must have at least five comparable firms (with positive bases of reference) within a given year to be included in the sample.

The stock price prediction equations for each of the nine models and the two-factor model are shortly described below. First, the prediction of firm *i*'s stock price using the arithmetic mean aggregated PE ratio ($\hat{PRICE}_{it}^{Arith,PE}$) is given by:

---

[213] Contrary to that, the derived two-factor model can also be applied if the reference variables are zero or negative. However, this model has other restrictions, which are outlined below.

[214] For this analysis, industries are defined as groups of certain NACE classes. The classification is largely identical with that of Sofka and Schmidt (2004). For a list of all industries and NACE classes included in this study, see Appendix 7.3.

$$\hat{PRICE}_{it}^{Arith,PE} = \left(\frac{1}{m_{it}} \cdot \sum_{j=1}^{m_{it}} \frac{PRICE_{jt}}{NI_{jt}} \right) \cdot NI_{it} \tag{5.7}$$

where $PRICE_{jt}$ is the stock price of firm $j \in CC_{it}$ at the last trading day in March of year t+1, CC_{it} is the set of comparable companies used for firm i for the year t, m_{it} is the number of companies in CC_{it}, and NI_{it} resp. NI_{jt} is the net income of firm i resp. firm j for the year t.

The prediction of firm i's stock price using the arithmetic mean aggregated PB ratio ($\hat{PRICE}_{it}^{Arith,PB}$) is expressed as:

$$\hat{PRICE}_{it}^{Arith,PB} = \left(\frac{1}{m_{it}} \cdot \sum_{j=1}^{m_{it}} \frac{PRICE_{jt}}{EQUT_{jt}} \right) \cdot EQUT_{it} \tag{5.8}$$

where $EQUT_{it}$ resp. $EQUT_{jt}$ the book value of equity of firm i resp. firm j for the year t. The other variables are as defined above.

The prediction of firm i's stock price using the arithmetic mean aggregated, equally-weighted PE/PB ratio ($\hat{PRICE}_{it}^{Arith,PEPB}$) is expressed as:

$$\hat{PRICE}_{it}^{Arith,PEPB} = \frac{1}{2} \cdot \left(\hat{PRICE}_{it}^{Arith,PE} + \hat{PRICE}_{it}^{Arith,PB} \right) \tag{5.9}$$

The prediction of firm i's stock price using the median aggregated PE ratio ($\hat{PRICE}_{it}^{Median,PE}$) is given by:

$$\hat{PRICE}_{it}^{Median,PE} = \left(median_j \left(\frac{PRICE_{jt}}{NI_{jt}} \right) \right) \cdot NI_{it} \tag{5.10}$$

where $median_j(x)$ means the median of x for all firms $j \in CC_{it}$.

The prediction of firm i's stock price using the median aggregated PB ratio ($\hat{PRICE}_{it}^{Median,PB}$) is calculated as:

$$\hat{PRICE}_{it}^{Median,PB} = \left(median_j \left(\frac{PRICE_{jt}}{EQUT_{jt}} \right) \right) \cdot EQUT_{it} \tag{5.11}$$

The prediction of firm i's stock price using the median aggregated, equally-weighted PE/PB ratio ($\hat{PRICE}_{it}^{Median,PEPB}$) is expressed as:

$$\hat{PRICE}_{it}^{Median,PEPB} = \frac{1}{2} \cdot \left(\hat{PRICE}_{it}^{Median,PE} + \hat{PRICE}_{it}^{Median,PB} \right) \quad (5.12)$$

The prediction of firm i's stock price using the harmonic mean aggregated PE ratio ($\hat{PRICE}_{it}^{Harmonic,PE}$) is computed as:

$$\hat{PRICE}_{it}^{Harmonic,PE} = \left(\frac{1}{m_{it}} \cdot \sum_{j=1}^{m_{it}} \left(\frac{PRICE_{jt}}{NI_{jt}} \right)^{-1} \right)^{-1} \cdot NI_{it} \quad (5.13)$$

The prediction of firm i's stock price using the harmonic mean aggregated PB ratio ($\hat{PRICE}_{it}^{Harmonic,PB}$) is given by:

$$\hat{PRICE}_{it}^{Harmonic,PB} = \left(\frac{1}{m_{it}} \cdot \sum_{j=1}^{m_{it}} \left(\frac{PRICE_{jt}}{EQUT_{jt}} \right)^{-1} \right)^{-1} \cdot EQUT_{it} \quad (5.14)$$

The prediction of firm i's stock price using the harmonic mean aggregated, equally-weighted PE/PB ratio ($\hat{PRICE}_{it}^{Harmonic,PEPB}$) is expressed as:

$$\hat{PRICE}_{it}^{Harmonic,PEPB} = \frac{1}{2} \cdot \left(\hat{PRICE}_{it}^{Harmonic,PE} + \hat{PRICE}_{it}^{Harmonic,PB} \right) \quad (5.15)$$

The derived two-factor model (equation 4.13) has to be slightly adjusted to be applicable in this empirical study. Because of lack of data, expected future values of accounting variables are replaced by current values. This, however, does not reduce the explanatory power of the analysis. The predicted price of the stock of firm i using the derived two-factor model can be expressed as follows:

$$\hat{PRICE}_{it}^{TWOFACTOR} = \left(1 + \frac{NI_{it}}{EQUT_{i(t-1)}} + d_{it} \cdot \Phi\left(\frac{d_{it}}{\sigma_{reo,t}} \right) + \sigma_{reo,t} \cdot \varphi\left(\frac{d_{it}}{\sigma_{reo,t}} \right) \right) \cdot EQUT_{it} \quad (5.16)$$

where $d_{it} = \hat{\alpha}_{it} - 1 + \left(\hat{\beta}_{it} - 1 \right) \cdot NI_{it} \cdot \left(EQUT_{i(t-1)} \right)^{-1}$

with $\hat{\alpha}_{it}$ and $\hat{\beta}_{it}$

being derived from a linear OLS-regression of those companies of CC_{it} that have $ROE \geq 0.05$ (with ROE_t defined as $NI_t \cdot \left(EQUT_{t-1}\right)^{-1}$),

$\sigma_{reo,t}$ is the five-year (unannualized) volatility of the CDAX calculated as $\sigma_{reo,t} = \sigma_{CDAX,t} \cdot \sqrt{5 \cdot 52}$, where $\sigma_{CDAX,t}$ is the historical standard deviation of the weekly market returns over a five year period ending at the last trading day in march of the year t+1,

$\Phi(x)$ denotes the cumulative standard normal distribution of x, and

$\varphi(x)$ denotes the standard normal distribution of x.

After the computation of the price prediction for each model, the next step is the calculation of the valuation errors. According to the majority of the literature on pricing accuracy (see e.g. Alford, 1992, Choudhary, 2004, Dittmann and Weiner, 2005) the absolute percentage valuation error $APVE_{it}$ for each of the ten models described in equations 5.7-5.16 is calculated as follows:

$$APVE_{it} = \left|\frac{\varepsilon_{it}}{PRICE_{it}}\right| = \left|\frac{PRICE_{it} - \hat{PRICE}_{it}}{PRICE_{it}}\right| \tag{5.17}$$

5.3.3 Results

Table 25 presents mean and median absolute percentage valuation errors, as well as the lower and the upper quartile and the standard deviation for each of the ten models. The valuation errors are pooled across the sample period to obtain more stable results. In order to reduce the impact of outliers, all predictions that result in an absolute valuation error of more than 100% were excluded from the calculation of the distribution measures. This is also a reasonable step from an economic perspective since appraisers in valuation practice would not believe in valuation results that lack economic plausibility.

Additionally, a Wilcoxon signed rank test – which is the non-parametric analogue to the paired t-test – is performed. The test allows conclusions about whether the valuation errors differ significantly between valuation models. The results of this test are displayed in Table 26. The grey marked fields indicate that the line-median exceeds the column-median.

Table 25. Absolute percentage valuation errors of different CCV models

Pricing models[a]	Observations	Lower quartile	Median	Mean	Upper quartile	Std. deviation
PE (Arithmetic)	268	0.191	0.374	0.423	0.620	0.276
PB (Arithmetic)	553	0.225	0.425	0.430	0.624	0.261
PEPB (Arithmetic)	290	0.154	0.325	0.371	0.544	0.256
PE (Median)	354	0.159	0.363	0.420	0.673	0.289
PB (Median)	693	0.195	0.404	0.421	0.622	0.259
PEPB (Median)	369	0.145	0.333	0.355	0.522	0.234
PE (Harmonic)	378	0.198	0.432	0.450	0.686	0.284
PB (Harmonic)	742	0.181	0.396	0.421	0.633	0.264
PEPB (Harmonic)	386	0.184	0.372	0.391	0.576	0.239
TWOFACTOR	422	0.170	0.356	0.390	0.608	0.260

[a] Absolute Percentage Valuation Error $APVE_{it} = \left| \frac{\varepsilon_{it}}{PRICE_{it}} \right| = \left| \frac{PRICE_{it} - \hat{PRICE}_{it}}{PRICE_{it}} \right|$,

where $\hat{PRICE}_{it}$ is calculated according to Equations 5.7-5.16.

[b] Pricing models are as defined in section 5.3.2.

The most striking result from this pricing accuracy analysis is that multi-factor models clearly outperform single-factor models, a result that is largely consistent with the findings of Cheng/McNamara (2000) and Beatty et al. (1999). The equally-weighted combined PE/PB ratios have the lowest median and mean *APVE* for each of the three aggregation methods. The derived two-factor model also exhibits higher valuation accuracy than the PE ratio or the PB ratio alone, no matter which aggregation mechanism is applied for the two single-factor models. Of course, the differences between the derived two-factor model and the arithmetic mean-aggregated PB ratio as well as the median-aggregated PE ratio are not significant, but that does not change the general statement. The better performance of multi-factor models is not surprising since a comparable company selection based on industry classification is by far more consistent with the principle of earnings/book value-based multi-factor models than with that of the two single-factor models.

An additional finding is that the derived two-factor model does not perform better than the equally-weighted combined PE/PB ratios. The pricing accuracy of the derived two-factor model is even significantly lower than that of the median-

Table 26. Wilcoxon signed rank test for equal valuation errors

Pricing Models[a,b]	PE Arith	PB Arith	PEPB Arith.	PE Med.	PB Med.	PEPB Med.	PE Harm	PB Harm.	PEPB Harm.	TWO FACTOR
PE Arith.	-	0.42	0.00 ***	0.36	0.01 ***	0.01 **	0.00 ***	0.01 ***	0.30	0.06 *
PB Arith.		-	0.00 ***	0.98	0.36	0.00 ***	0.08 *	0.15	0.62	0.82
PEPB Arith.			-	0.01 **	0.00 ***	0.69	0.00 ***	0.00 ***	0.00 ***	0.18
PE Med.				-	0.06 *	0.00 ***	0.00 ***	0.02 **	0.59	0.15
PB Med.					-	0.00 ***	0.97	0.20	0.00 ***	0.09 *
PEPB Med.						-	0.00 ***	0.00 ***	0.00 ***	0.07 *
PE Harm.							-	0.59	0.00 ***	0.00 ***
PB Harm.								-	0.00 ***	0.01 ***
PEPB Harm.									-	0.17

[a] p-values are reported; * (**, ***): significant at the two-tailed 10% (5%, 1%) level; grey marked fields indicate that the line-median exceeds the column-median.
[b] Pricing models are as defined in section 5.7-5.16.

aggregated equally-weighted PE/PB ratio. However, with respect to the other two combined PE/PB ratios the results are ambiguous: The arithmetic mean-aggregated PE/PB ratio has a higher pricing accuracy while the harmonic mean-aggregated PE/PB ratio has a lower pricing accuracy, but both differences are not significant.

Apart from these findings no clear results can be derived from the examinations. The PB ratio performs sometimes better and sometimes worse than the PE ratio, depending on the aggregation mechanism applied. Moreover, none of the aggregation methods demonstrated significantly more accurate valuation results.

The key lessons from this analysis are the following: Multi-factor models based on book value and earnings outperform single-factor models, such as the PE and the PB ratio, if the set of comparable companies is selected based on industry classification. This is principally a strong argument for the two-factor model that was derived in section 4.3.2. However, a comparison of the multi-factor models reveals that the derived two-factor model cannot outperform simple equally weighted combined PE/PB ratios. This latter finding suggests that there is still room for improvement of the two-factor model.[215] The fact that no general statement about the favourability of one of the two single-factor models – the PE ratio and the PB ratio – can be made, emphasizes the finding in section 5.2 that an assessment about the valuation power of these two ratios is only possible if one considers the specific valuation circumstances.

215 See section 4.3.2.5 for a discussion about the shortcomings of the two-factor model.

6 Concluding Remarks

The valuation of a company based on how similar companies are priced on the stock market is called *the market approach to comparable company valuation* (CCV). CCV is one of the most controversial approaches to business valuation. While it enjoys widespread popularity in valuation practice, most academics criticize it because of alleged lack of theoretical foundation and oversimplification of the valuation procedure. This dispute is to a large extent due to a broad misunderstanding regarding the economic implications of different CCV models as well as regarding how capital market theory fits into the CCV framework. As a result, this valuation approach is often misused in practice, and most theoreticians stick to technical details while loosing sight of the big picture: the determination of a corporate value. The present study makes two contributions to the public discussion about CCV:

(1) It adds to fill the gap between academics and practitioners *on a theoretical basis*. This is done in chapter 2 by describing how CCV fits into the whole company valuation framework. In this context, several value theories are presented, the differences – but also the similarities – between market price and corporate value are discussed, a comparison with other valuation approaches is performed, and some special aspects of the CCV process (such as the need for high-quality accounting variables, the way reported variables can be adjusted, the aggregation of the variables if more than one comparable company is used, and the consideration of premiums and discounts) are pointed out.

In chapter 3, the two main tasks of CCV – as well as their determinants – are discussed in-depth: the comparable company selection and the valuation model choice. The first task, comparable company selection, is a function of the similarity of companies and of the efficiency and pricing accuracy of the stock market. Several similarity criteria are presented for different valuation models and a broad overview of informational capital market efficiency is given. The second task, valuation model choice, is a function of the value relevance of the applied reference variables (the accounting figures that link the price of the comparable companies to the target company), the future similarity between comparable companies and the target company, as well as technical restrictions inherent in certain valuation models. At the end of this chapter it is shown that both tasks are in no way independent from each other, but are closely connected and even interdigitate. Finally a "solution package" for sound CCV is given.

Chapter 4 contains a short overview of the valuation process of immediate CCV (i.e. CCV without a linking factor) and single-factor CCV (i.e. CCV with one linking factor – the price-earnings ratio is an example of single-factor CCV).

Part of this overview is a presentation of common mistakes that are made when performing single-factor CCV. Additionally, some light is shed on the biggest challenges that analysts and investors face when applying single-factor models. After that, some existing multi-factor models (i.e. models that use more than one linking factor) are explained. The main part of chapter 4, however, focuses on the derivation of a two-factor CCV model based on the book value of equity and earnings. From a theoretical point of view, this model largely overcomes the problems associated with single-factor models. Additionally, it is applicable in practice.

(2) Guided by financial and valuation theory, this study explores several specific topics of CCV on *an empirical basis*. On the one hand, this is done via three different surveys amongst financial analysts and institutional investors, each conducted as part of the monthly Financial Market Survey performed by the Centre for European Economic Research (ZEW), Mannheim. These surveys concern the following topics: application of CCV in general, capital market efficiency in 2005, and the determination of long-term earnings' growth rates. On the other hand, by using a large and quite representative sample of non-financial German companies that publish consolidated annual reports according to IAS/IFRS or US-GAAP and that were listed at the German Stock Exchange during the period 1998-2003, several hypotheses about value relevance of accounting figures are tested (chapter 5). High value relevance is appreciable since this is the basis of a powerful valuation model and, thus, ensures high valuation accuracy. Various methodologies are applied here: A simple comparison of coefficients of determinationis followed by more sophisticated regression models. Finally, an innovative approach is applied in order to overcome a selection bias associated with two of the hypotheses. This innovative approach is called the *matching estimator* approach and its basic proceeding is the following: Point of origin is the situation where coefficients should be tested for significant differences between two heterogeneous groups of observations. In order to better deal with this heterogeneity, the matching estimator approach alters the nature of one of these two groups such that this group is virtually identical to the other group with respect to some ex ante defined variables. Assuming that this set of ex ante defined variables covers most of the initial differences between both groups while it does not include the variable under consideration (the variable for which the coefficients should be compared between both groups), then any remaining differences can now be assigned to the variable under consideration. In addition to the value relevance study, an examination of the pricing accuracy of the two-factor model that was derived in chapter 4 is conducted.

6.1 Implications for Business Valuation

Despite the fact that formal theoretical derivations and empirical analyses make up a big part of the whole study, the focus is clearly on economic implications. The most important findings with respect to business valuation are listed below.

It has been discussed theoretically that:

- The predominant aim of CCV is to assign potential market prices to a company. In this context, it is important to note that the terms "value" and "price" are not identical. However, under certain circumstances the market price of stocks can be seen as a substitute for the value of stocks. Whether this is the case depends on the specific valuation situation, the purpose of valuation and on how developed the stock market is.
- CCV and direct valuation approaches (such as the discounted cash flow method) are very similar in their methodology. The big difference is that direct valuation approaches rely on individual forecasts of future cash flows where CCV relies on the market implied forecasts of these cash flows. The appraiser, however, still needs forecasting skills when performing CCV. The forecasting problem is shifted from explicit estimates of future business development of single companies (in the case of direct valuation) to the prediction of future similarity between the target company and the comparable companies. It is also highly important to note that the problems of investment comparability that are associated with direct valuation approaches are at least as severe as those associated with CCV – especially if the Capital Asset Pricing Model (CAPM) serves as a tool to determine the discount rate in direct valuations. This becomes clear when looking at the comparable investment in both cases. In CCV it consists of (existing) similar, publicly traded companies, while in direct approaches using the CAPM it consists of a combination of the riskless asset and the beta weighted (unobservable) market portfolio. Moreover, in the latter case the validity of the neo-classical view on capital markets must be assumed.
- When determining the number of comparable companies, appraisers have to consider that the improvement of the pricing accuracy due to a higher number of companies usually comes at the cost of a decrease of the peer group quality due to potential imperfect substitutes. Similarly, appraisers that choose a lower number of comparable companies usually face fewer peer group quality risks, but the risk of pricing inaccuracies is higher. This dilemma is a major problem in the process of selecting comparable companies.
- In order to accurately perform CCV, the comparable companies have to be similar to the target company in many aspects. The pure reliance on "industry classification" as a comparability criterion – as is often done in valuation practice – is typically insufficient. This gives rise to the suggestion that there is a lack of adequate comparable companies in many valuation cases.
- An important criterion for the usefulness of different CCV models is the value relevance of the reference variables (i.e. the accounting figures that serve as a linking tool). In this context, value relevance means how well the reference variables explain stock prices and, thus, how well the reference variables are an indicator of future cash flows of the target company.
- Enterprise CCV (i.e. valuation of the company as a whole) is – contrary to a popular view – not an appropriate approach to compare companies with permanently different capital structures. However, if these differences in capital struc-

tures are only temporary in nature, then enterprise CCV models might indeed yield more accurate valuation results than equity CCV models.

- The price-sales ratio is – despite its popularity in valuation practice – not backed by financial theory since the numerator is an equity value and the denominator is an enterprise value. This offends against the principle of consistent definition of valuation ratios, which is one of the underlying concepts in CCV.
- The price-earnings-to-growth (PEG) ratio aims at eliminating the growth rate of financial benefits as a similarity criterion for the peer group and, thus, at comparing companies with different expected growth rates. However, it fails to do so. In fact, a sensitivity analysis shows that for high-growth company valuation the PEG ratio reduces the influence of the growth rate on the comparable company selection but does not delete it. Contrarily, when valuing low growth companies by using the PEG ratio, the influence of the growth rate on peer group selection is even aggravated.
- CCV is not a magic bullet. It has to be thoroughly analysed case-by-case whether it can be applied or not.

It has been shown both theoretically and empirically that:

- The informational efficiency of the stock market is quite high. Not only many academic research studies support this view, but also a survey amongst financial analysts and institutional investors, which was conducted in order to figure out the state of informational market efficiency in Germany in 2005, emphasises this. However, markets are still far away from perfect efficiency and, thus, analysts have to consider the existence of some anomalies when performing CCV. In this context, it is important to note that CCV might still be a reasonable valuation approach even if markets are not perfectly efficient, as long as the amount of information processed in direct valuation approaches is not superior to the information inherent in market prices.
- The determination of the long-term growth rate of earnings is one of the most difficult tasks in business valuation. The results from a survey amongst financial analysts and institutional investors emphasize this. However, the proper performance of single-factor CCV models requires a set of comparable companies that has growth rates identical to the target. This dilemma clearly reduces the power of single-factor CCV models and calls for the derivation of new models that do not (or at least: not that much) rely on the growth rate as a comparability criterion.
- Accrual-based financials (such as earnings) are on average slightly more value relevant than cash-based financials (cash flows). This means that accrual-based multiples are expected to yield more accurate valuation results than cash flow multiples (assuming that there is no lack of comparable companies). This is not – as sometimes mentioned – a contradiction to valuation principles, but rather a sign for the suitability of accrual accounting under IAS/IFRS resp. US-GAAP as a means of properly allocating future expected cash flows to the current period.

- Free cash flows – both, to equity and to the firm – have very low value relevance. This finding supports the hypothesis that current free cash flow is a bad predictor of future cash flows. In fact, it is not possible to differentiate between a low free cash flow due to a low operating cash flow on the one hand (which is a bad sign) and due to high investments on the other hand (which is usually a good sign because of the future expected benefits of these investments). As a consequence, single free cash flows are not very meaningful and, thus, free cash flow multiples cannot be regarded as an appropriate valuation model.
- Sales are more value relevant than EBIT and of similar value relevance as EBITDA. This result is somewhat surprising since sales are expected to have the lowest value relevance among these three variables (because of the non-consideration of the cost of goods sold). One explanation for this empirical finding is that sales multiples enjoyed a widespread popularity during the high-tech boom and, thus, partly "drove" market prices.
- A two-factor model that is based on the book value of equity and earnings, which considers a company's option to abandon business or to reorganise, has some explanatory power for the German stock market. This is particularly emphasised by two empirical findings. First, it has been found that the shape of the model is as predicted (i.e. the price-book ratio is a convex function of the return on equity). Second, if the comparable companies are selected based on industry classification, the pricing accuracy of this model is higher than the accuracy of the price-earnings ratio or the price book ratio – no matter which aggregation method is applied. However, the derived two-factor cannot outperform a simple equally weighted price-earnings/price-book ratio model.
- From a theoretical point of view, the major advantage of the two-factor model is that the long-term growth rate of financial benefits is no longer a comparability criterion. In fact, the criterion "growth rate" is substituted here by the "persistence of abnormal earnings". Persistence, in turn, can be assumed to be equal for all companies of one industry. Therefore, the two-factor model can be seen as an improvement to classical single-factor models in that it has weaker similarity requirements. Additionally, the model can conclusively assign positive stock prices to currently negatively performing companies, and allows depicting expected future earnings development in an economicly sounder way than single-factor models (due to its methodological similarity to the Ohlson model). Furthermore, the model gives advice how earnings multiples and book value multiples can be combined. Nevertheless, its application is complicated and, thus, it cannot be expected that this model will become very popular amongst analysts. Particularly, the determination of the time to expiration of the embedded option to reorganise the company is associated with problems.
- Book value of equity and earnings have joint value relevance that is higher than the value relevance of each of these variables alone. Moreover, a simple equally weighted price-earnings/price-book-ratio model clearly outperforms the price-earnings ratio and the price-book ratio in terms of pricing accuracy if the comparable companies are selected based on industry classification. Thus, there is strong evidence that the joint application of book value multiples and earnings multiples increases valuation quality.

- The value relevance of earnings is reduced and the value relevance of book value is increased for firms that operate in highly competitive industries compared to firms that do not. This suggests that book value multiples yield more accurate valuation results when valuing companies in highly competitive industries. Likewise, earnings multiples are expected to yield more accurate valuation results when valuing companies in non-competitive industries.
- The value relevance of book value is slightly increased for asset heavy (i.e. capital intensive) firms compared to firms that are not asset heavy. This suggests that book value multiples yield more accurate valuation results when valuing asset heavy companies than when valuing other companies.
- The value relevance of earnings is reduced for firms that have high earnings-adjusted creditworthiness compared to firms that are of poor earnings-adjusted creditworthiness. This suggests that earnings multiples yield more accurate valuation results when valuing a low-rated company than when valuing a highly creditworthy company, assuming both companies have the same level of current profitability.
- In general, companies trade at lower multiples if they operate in highly competitive industries than in non-competitive industries. This is the case because industry forces drive future earnings to decline over time in competitive industries, which subsequently results in lower relative stock prices.
- In general, companies trade at higher multiples if they have high creditworthiness than if they are of low creditworthiness. This is the case because a high rating implies less financial risk, which subsequently results in higher relative stock prices.

6.2 Implications for Future Research

The findings of this study have implications for future theoretical and empirical studies about CCV. They can be summarised as follows:

- Research about CCV can be conducted from many different perspectives. In fact it combines elements of accounting research, of finance but also of corporate strategy. Hitherto, many studies focus on just one of these branches of CCV research, and mask out the other ones. To overcome problems of biased conclusions, future research should rather follow an integrated, holistic approach. This will help to properly assess CCV as a valuation approach.
- Multi-factor models have been extensively discussed in finance literature as a means of determining expected returns (both types: models that are derived from an equilibrium theory and purely empirical models). The standing of multi-factor models within the scope of CCV is, however, still very weak. Of course, practitioners apply multi-model approaches, but thus far there is no guidance on how to weight the results from each single model. To put it more precisely, the determinants of interaction of single-factor models are highly unknown territory. Moreover, hitherto there are no (properly) theoretically derived multi-factor models other than those that build on the Ohlson (1995)

model. Future research should deal with that topic both on a theoretical and an empirical basis.

- There are several studies about CCV in general, for both, valuation model choice and peer group selection. However, there is a lack of empirical research that is conducted for specific valuation circumstances. In real valuation settings, there is no "one size fits all" model and, thus, one model might work very well in a certain situation, but might totally fail to accurately assign a value in other situations. In order to provide practice-oriented research in the future, it is highly important to provide concrete guidance on when to apply which model for which companies. There are so many different valuation situations and the present study could shed light on but a few (valuation dependent on the degree of competitiveness of the industry, dependent on the creditworthiness of the company, and dependent on the asset heaviness of the company). Future research should without doubt address this issue.
- There is a need for high quality data in order to properly assess the pricing accuracy of CCV models. In particular, this means that a longer sample period is necessary for future research in order to gain more observations for companies that publish financial statements according to IAS/IFRS or US-GAAP in Germany – and especially to gain more variations over time. This will allow applying fixed-effects models in order to better take into account the unobserved part of firm heterogeneity. Additionally, if the focus is on valuation model choice, future research should use specified sets of comparable companies that are provided by investment banks. This will help turn the attention to the real points of interest. However, a major obstacle for CCV research is that these data are not readily available thus far. Hopefully, this will change in the near future. Finally, the accurate measurement of variables becomes a crucial task, especially if certain valuation circumstances have to be modelled.
- The principle of the *matching estimator approach* can probably serve as a tool to select the comparable companies. Future research should take that possibility into consideration.

7 Appendix

7.1 Detailed Derivation of the Two-Factor Comparable Company Valuation Model

To establish the functional relationship between the price scaled by book value of equity (PB) and the expected future return on equity (ROE), the expected maximum of the reorganisation value and the recursion value has to be computed:

$$PB = E\left[\max\left(V_{reo}, V_{rec}\right)\right]$$

$$PB = E\left[\max\left\{1 + ROE + \varepsilon, \alpha + \beta \cdot ROE\right\}\right]$$

$$PB = \int_{-\infty}^{+\infty} \max\left\{1 + ROE + \varepsilon, \alpha + \beta \cdot ROE\right\} f_{\varepsilon}(\varepsilon) d\varepsilon$$

where ε is the normally distributed additive error with expectation value E[ε]=0 and standard deviation $\sigma[\varepsilon]=\sigma_{reo}$ ($\varepsilon \sim N(0, \sigma_{reo})$) and $f_{\varepsilon}(\varepsilon)$ denotes its probability density function.

$$PB = \int_{-\infty}^{+\infty} \left[\left(\alpha + \beta \cdot ROE\right) \cdot 1_{\left[-\infty, d\right]}\left(\varepsilon\right) + \left(1 + ROE + \varepsilon\right) \cdot 1_{\left[d, +\infty\right]}\left(\varepsilon\right)\right] \cdot f_{\varepsilon}(\varepsilon) d\varepsilon$$

where $1_{\left[a, b\right]}\left(\varepsilon\right) := \begin{cases} 1, \text{ if } \varepsilon \in \left[a, b\right] \\ 0, \text{ if } \varepsilon \notin \left[a, b\right] \end{cases}$

and the parameter d is defined as the realisation of ε which leads to investors' indifference between recursion value and reorganisation value. It is obtained by setting V_{reo} equal to V_{rec} and solving for ε which yields

$$d \equiv V_{rec} - E\left[V_{reo}\right] = \alpha - 1 + \left(\beta - 1\right) \cdot ROE$$

$$PB = \int_{-\infty}^{d} \left(\alpha + \beta \cdot ROE\right) \cdot f_{\varepsilon}(\varepsilon) d\varepsilon + \int_{d}^{+\infty} \left(1 + ROE + \varepsilon\right) \cdot f_{\varepsilon}(\varepsilon) d\varepsilon$$

$$PB = (\alpha + \beta \cdot ROE) \int_{-\infty}^{d} f_\varepsilon(\varepsilon) d\varepsilon + (1 + ROE) \int_{d}^{+\infty} f_\varepsilon(\varepsilon) d\varepsilon + \int_{d}^{+\infty} \varepsilon \cdot f_\varepsilon(\varepsilon) d\varepsilon$$

$$PB = (\alpha + \beta \cdot ROE) \int_{-\infty}^{d} f_\varepsilon(\varepsilon) d\varepsilon + (1 + ROE) \cdot \left(1 - \int_{-\infty}^{d} f_\varepsilon(\varepsilon) d\varepsilon\right) + \int_{d}^{+\infty} \varepsilon \cdot f_\varepsilon(\varepsilon) d\varepsilon$$

$$PB = (1 + ROE) + (\alpha - 1 + (\beta - 1) ROE) \int_{-\infty}^{d} f_\varepsilon(\varepsilon) d\varepsilon + \int_{d}^{+\infty} \varepsilon \cdot f_\varepsilon(\varepsilon) d\varepsilon$$

Since $\int_{-\infty}^{d} f_x(x) dx = \Phi\left(\frac{d}{\sigma}\right)$ and $\int_{d}^{+\infty} \varepsilon \cdot f_\varepsilon(\varepsilon) d\varepsilon = \sigma \cdot \varphi\left(\frac{d}{\sigma}\right)$ with $\Phi(x)$ denoting the cumulative standard normal distribution of x, and $\varphi(x)$ denoting the standard normal distribution of x, the final function is expressed as:

$$PB = E[V_{reo}] + d \cdot \Phi\left(\frac{d}{\sigma_{reo}}\right) + \sigma_{reo} \cdot \varphi\left(\frac{d}{\sigma_{reo}}\right).$$

7.2 Proof of the Convergence of R²

The following is proof of the convergence of R^2 to 1 if the same random variable is added to the dependent and the independent variable in a simple linear regression model when using the ordinary least square (OLS) method, and if this added variable approaches infinity.

In a regression equation where the stochastic variable z_i, $i \in I := \{1,\ldots,m\}$ is added to both the endogenous and the exogenous variable, i.e.:

$$(y_i + z_i) = b_0 + b_1 \cdot (x_i + z_i) + \varepsilon_i$$

with x_i, y_i, $i \in I$ being stochastic and bound to finite terms, the coefficient of determination R^2 is defined as:

$$R^2 = 1 - \frac{\sum_{i=1}^{n} \left((y_i + z_i) - \hat{b}_0 - \hat{b}_1 \cdot (x_i + z_i)\right)^2}{\sum_{i=1}^{n} \left((y_i + z_i) - \frac{1}{n} \cdot \sum_{i=1}^{n} (y_i + z_i)\right)^2}$$

The regression coefficients are estimated as:

$$\hat{b}1 = \frac{\sum_{i=1}^{n}\left(y_i + z_i - \overline{(y+z)}\right)\left(x_i + z_i - \overline{(x+z)}\right)}{\sum_{i=1}^{n}\left(y_i + z_i - \overline{(y+z)}\right)^2} = \frac{\operatorname{cov}\left(\left(y_i + z_i\right),\left(x_i + z_i\right)\right)}{\operatorname{var}\left(y_i + z_i\right)}$$

and

$$\hat{b}0 = \frac{1}{n}\sum_{i=1}^{n}\left(y_i + z_i\right) - \hat{b}1 \cdot \frac{1}{n}\sum_{i=1}^{n}\left(x_i + z_i\right) \quad ,$$

with $\overline{(x+z)}$ and $\overline{(y+z)}$ denoting the arithmetic means of $x_i + z_i \;\; , i \in I$ and $y_i + z_i \;\; , i \in I$, respectively.

If $z_i \to \infty$, then

$$\lim_{z_i \to \infty}\left[\operatorname{cov}\left(\left(y_i + z_i\right),\left(x_i + z_i\right)\right)\right] = \operatorname{var}(z_i) \quad ,\forall i$$

and

$$\lim_{z_i \to \infty}\left[\operatorname{var}\left(y_i + z_i\right)\right] = \operatorname{var}(z_i) \quad ,\forall i$$

and thus

$$\lim_{\substack{z_i \to \infty \\ \forall i \in I}}\left[\hat{b}1\right] = 1 \quad ;$$

Substituting $\hat{b}1 = 1$ into the equation that defines $\hat{b}_0$ yields:

$$\lim_{\substack{z_i \to \infty \\ \forall i \in I}}\left[\hat{b}0\right] = 0 \quad .$$

Since $\hat{b}_1$ approaches 1 and consequently $\hat{b}_0$ approaches 0 for $z_i \to \infty$, the numerator of the fraction in the R^2-equation approaches 0 as z_i gets closer to infinity, and the denominator is positive as long as there is variation in z_i. Thus, the whole fraction approaches 0 and consequently R^2 approaches 1:

$$\lim_{\substack{z_i \to \infty \\ \forall i \in I}}\left[R^2\right] = 1$$

7.3 Industries Included in the Empirical Study

Industry number	Industry name	NACE Codes
1	Manufacture of motor vehicles	34, 35
2	Medical, precision, and optical instrument	33
3	Manufacture of machinery and equipment	29
4	Chemicals/petroleum/life science	23, 24, 85
5	Plastic/rubber	25
6	Manufacture of electrical machinery	30, 31, 32
7	Glass/ceramics	26
8	Wood/paper/publishing	20, 21, 22
9	Metal	27, 28
10	Textiles and leather	17, 18, 19
11	Manufacture of furniture, jewellery, sports equipment and toys	36, 37
12	Food and tobacco	15, 16
13	Wholesale trade	51
14	Retail and motor trade	50, 52
15	Transportation and communication	60, 61, 62, 63, 64.1
16	Financial intermediation	65, 66, 67
17	ICT services	72, 64.2
18	Technical services	73, 74.2, 74.3, 74.5, 93
19	Consulting	74.1, 74.4
20	Real estate activities and renting	70, 71
21	Sporting and other business activities	92, 74.8
22	Utilities	11, 40
23	Construction	45

7.4 Descriptive Statistics

7.4.1 Descriptive Statistics by Creditworthiness

Mean (n=675) Standard deviation in italics	Original rating classes of Creditreform[b]					
Variables[a]	1	2	3	4	5	6
PRICE	31.8	19.1	8.8	5.8	-	1.4
	21.5	20.4	12.4	7.6		1.2
EV	89.3	46.7	24.1	9.7	-	2.4
	78.2	63.6	41.9	9.1		2.0
NI	2.2	0.9	-1.1	-1.4	-	-2.3
	3.1	2.3	4.0	2.3		2.0
EBIT	4.2	1.6	-0.9	-1.1	-	-2.0
	6.2	3.6	4.0	2.1		1.9
EBITDA	8.1	3.7	1.3	-0.2	-	-1.2
	9.7	7.0	7.3	1.7		0.9
SALES	73.3	45.3	22.7	9.1	-	1.1
	74.0	81.2	55.1	15.5		1.0
FCFE	8.1	6.4	3.2	1.5	-	1.5
	13.1	14.0	18.2	4.4		4.9
FCFF	9.5	5.8	2.3	0.4	-	-0.6
	12.9	12.3	9.7	3.1		1.3
CFO	5.3	2.9	0.5	-0.5	-	-1.3
	6.4	6.5	2.5	2.2		1.0
EQUT	20.2	12.3	7.4	5.0	-	2.0
	15.2	14.0	11.0	4.3		3.0
TA	77.7	40.0	22.7	8.9	-	2.9
	83.6	63.4	45.2	4.3		3.6
# Observations	101	270	230	66	0	8

[a] Variables are as defined in section 5.2.1.
[b] An overview of the Creditreform rating system can be found in section 1.2.3.2.

7.4.2 Descriptive Statistics by Industry Structure

Mean (n=810) Standard deviation in italics	Characterisation of industries[b]		
Variables[a]	Competitive	Neutral	Not competitve
PRICE	10.7	15.7	27.4
	14.4	*18.7*	*22.9*
EV	24.8	33.5	82.4
	44.5	*56.1*	*83.5*
NI	-0.5	0.1	1.7
	3.3	*2.4*	*3.8*
EBIT	-0.1	0.7	3.3
	3.5	*4.0*	*7.0*
EBITDA	1.7	2.1	7.1
	6.8	*6.1*	*10.8*
SALES	22.7	33.0	68.7
	50.5	*91.8*	*93.7*
FCFE	4.2	3.2	8.5
	16.0	*7.5*	*15.3*
FCFF	3.3	3.4	9.5
	9.9	*10.2*	*15.7*
CFO	1.1	1.8	5.1
	4.1	*7.3*	*8.0*
EQUT	7.3	8.7	20.7
	10.3	*20.5*	*17.5*
TA	21.3	26.5	75.8
	45.2	*67.1*	*68.2*
# Observations	438	234	138

[a] Variables are as defined in section 5.2.1.
[b] Details about industry classification can be found in section 1.2.3.2 and in Appendix 7.3 .

7.4.3 Descriptive Statistics by Asset Heaviness

Mean (n=928) Standard deviation in *italics*		Nature of company[b]	
		Asset heavy	Not asset heavy
Variables[a]	PRICE	14.6 *17.4*	14.4 *18.1*
	EV	40.8 *61.5*	33.8 *57.3*
	NI	-0.1 *3.2*	0.4 *3.4*
	EBIT	0.6 *4.6*	0.6 *4.4*
	EBITDA	3.0 *6.9*	2.4 *7.6*
	SALES	27.6 *50.2*	38.5 *91.5*
	FCFE	6.6 *18.2*	2.7 *6.1*
	FCFF	5.4 *12.7*	3.0 *9.1*
	CFO	2.1 *5.3*	1.5 *6.3*
	EQUT	11.9 *14.8*	8.4 *16.0*
	TA	38.0 *63.3*	27.8 *62.5*
# Observations		440	488

[a] Variables are as defined in section 5.2.1.
[b] Details about company classification can be found in section 1.2.3.2.

7.5 Annual R² for Single-Factor Models

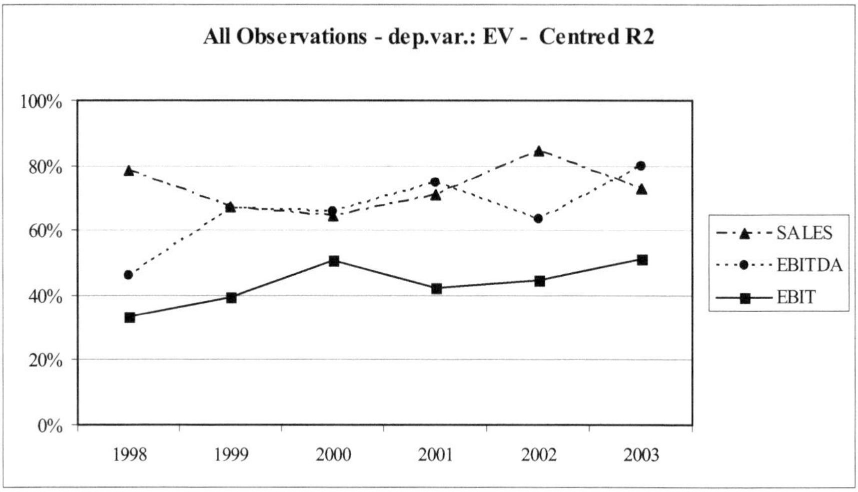

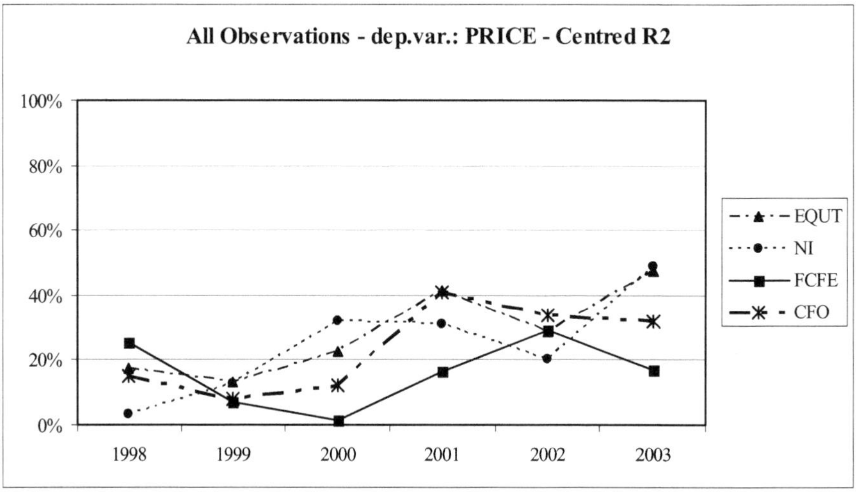

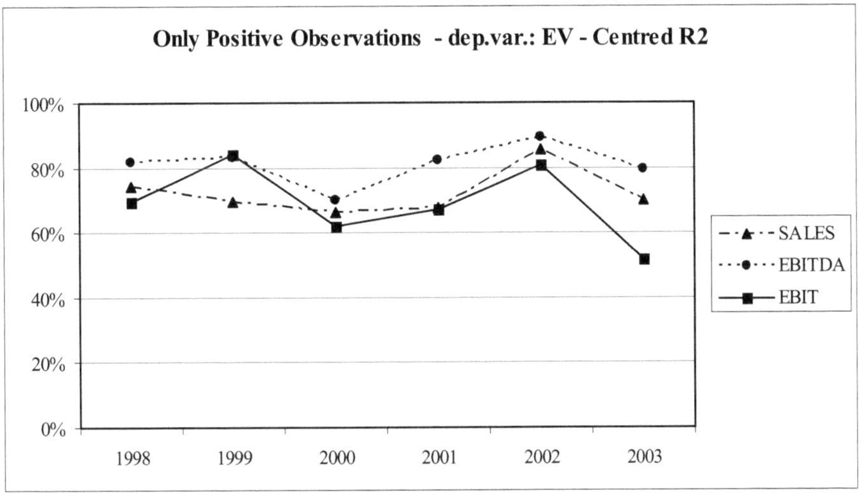

Only Positive Observations - dep.var.: EV - Centred R2
100%
80%
60%
40%
20%
0%
1998
1999
2000
2001
2002
2003
SALES
EBITDA
EBIT

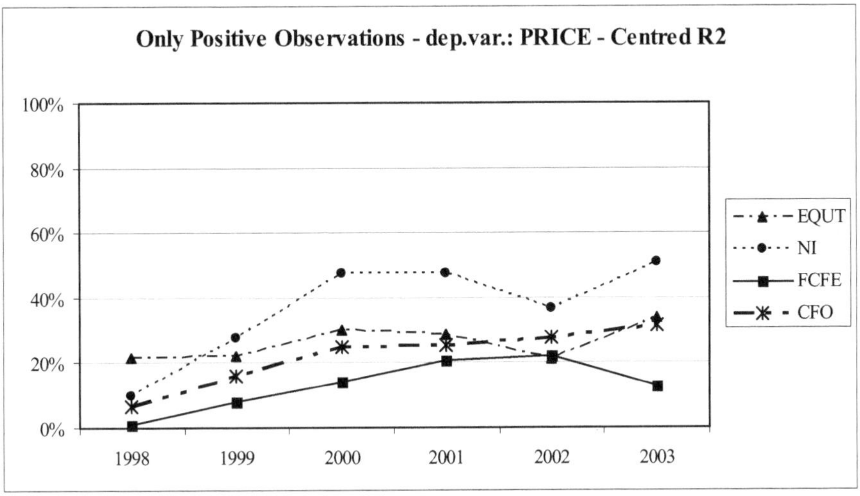

Only Positive Observations - dep.var.: PRICE - Centred R2
100%
80%
60%
40%
20%
0%
1998
1999
2000
2001
2002
2003
EQUT
NI
FCFE
CFO

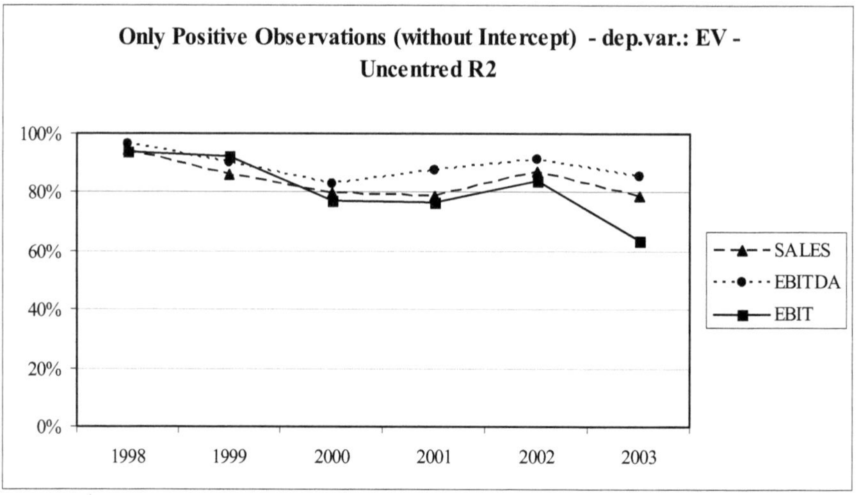
Only Positive Observations (without Intercept) - dep.var.: EV - Uncentred R2
100%
80%
60%
40%
20%
0%
1998
1999
2000
2001
2002
2003
SALES
EBITDA
EBIT

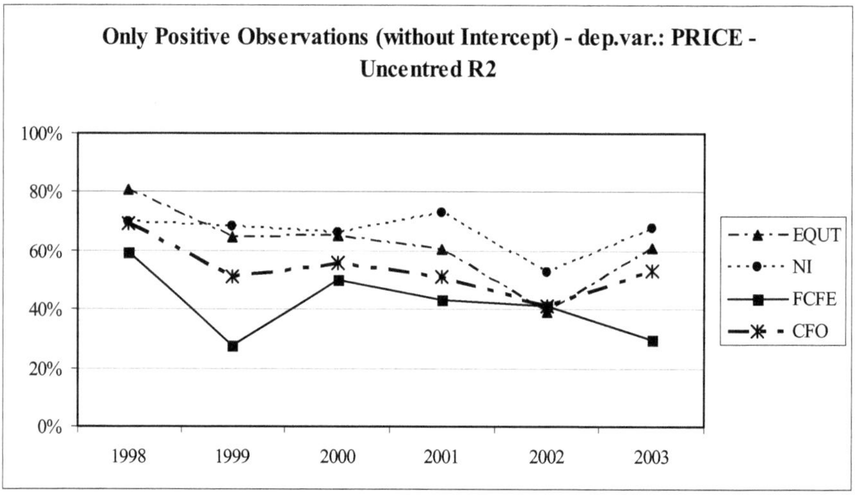
Only Positive Observations (without Intercept) - dep.var.: PRICE - Uncentred R2
100%
80%
60%
40%
20%
0%
1998
1999
2000
2001
2002
2003
EQUT
NI
FCFE
CFO

7.6 Adjusting Creditworthiness for the Impact of Earnings

In order to overcome some of the problems that are potentially associated with the use of the ordinally scaled (strictly positive) dependent variable *CRW* (section 5.2.3.2), a Box-Cox Transformation (Box/Cox 1964) is applied here as a tool to transform this dependent variable into a new variable such that this new variable is approximately normally distributed. This transformation probably better allows fulfilling the OLS-regression requirements.

The transformation functional according to the Box-Cox approach has the following general form:

$$CRW^{(\lambda)} = \frac{CRW^{\lambda} - 1}{\lambda}$$

$$\text{with } CRW^{(\lambda=0)} = ln(CRW)$$

The main contribution of the Box-Cox approach is that the parameter λ and the regression coefficients themselves could be estimated simultaneously using the method of *maximum likelihood.* Building on the concept presented in Table 6, the following regression equation is estimated:

$$CRW_{it}^{(\lambda)} = b_1 \cdot \left(\frac{NI_{it}}{PRICE_{it}} \right) + \varepsilon_{it}$$

which yields the following results:

	Coefficient	z-value
$\frac{NI}{PRICE}$	-632.17	-
λ	1.57	38.2***

The association between *CRW* and $CRW^{(\lambda)}$ is illustrated in the figure below. For convenience, some graphs for classical values of λ other than 1.57 are also plotted.

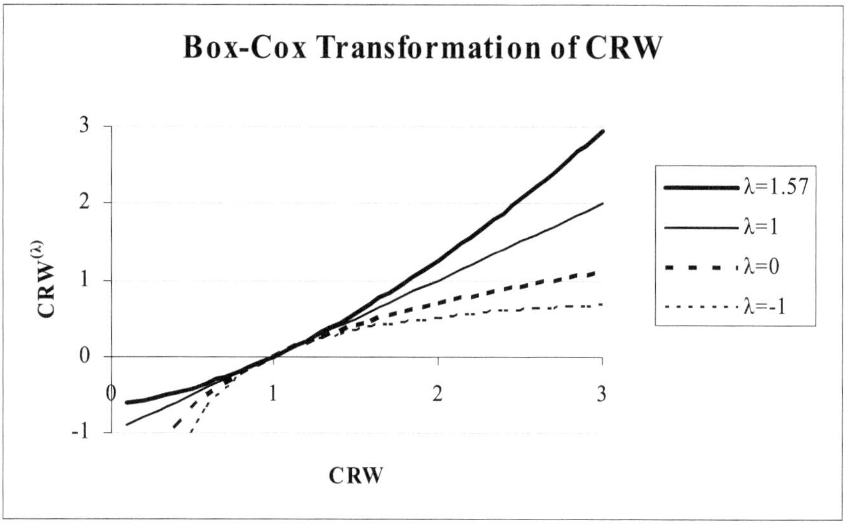

Source: ZEW

The earnings adjusted creditworthiness using the Box-Cox transformation ($ACRW^{BOX}$) equals the residuals of the former regression model. It can be calculated as follows:

$$ACRW_{it}^{BOX} = \widehat{\varepsilon_{it}} = CRW_{it}^{(\lambda)} - \widehat{b_1} \cdot \left(\frac{NI_{it}}{PRICE_{it}} \right) = \frac{\left(CRW_{it}\right)^{1.57} - 1}{1.57} + 632.17 \cdot \left(\frac{NI_{it}}{PRICE_{it}} \right)$$

The variable ACRW^{BOX} is now used to construct the dummy variable D4^{BOX}. In order to maintain consistency with the dummy variable D3, the cut-off rate to divide the sample into companies that are of "high earnings adjusted creditworthiness" and those that are not, is set such that the percentage of companies in both groups remain the same as under the dummy D3. Consequently, the new cut-off rate is at ACRW^{BOX}=1,200.

Thus, the dummy rule is:

$\mathbf{D4}^{\mathbf{BOX}}$ = 1 if a company has a $\text{ACRW}^{BOX} < 1,200$ (i.e. high earnings adjusted creditworthiness)

$\mathbf{D4}^{\mathbf{BOX}}$ = 0 if a company has a $\text{ACRW}^{BOX} \geq 1,200$ (i.e. no high earnings adjusted creditworthiness)

Impact of box-cox earnings adjusted creditworthiness on the relative valuation roles of book value and earnings

Regression models (yearly dummies are not explicitly listed)

$$P_{it} = b_0 + b_1 \cdot EQUT_{it} + b_2 \cdot NI_{it} + \varepsilon_{it} \quad \text{(A)}$$

sub-sample:high adj. creditworthiness

$$P_{it} = b_0 + b_1 \cdot EQUT_{it} + b_2 \cdot NI_{it} + \varepsilon_{it} \quad \text{(B)}$$

sub-sample: low adj. creditworthiness

$$P_{it} = b_0 + b_1 \cdot D4_{it}^{BOX} + b_2 \cdot EQUT_{it} + b_3 \cdot D4_{it}^{BOX} \cdot EQUT_{it} + + b_4 \cdot NI_{it} + b_5 \cdot D4_{it}^{BOX} \cdot NI_{it} + \varepsilon_{it} \quad \text{(C)}$$

Regression results [a,b]

	Equation		
Variables[c]	(A)	(B)	(C)
Cook-Weisberg	***	***	***
Intercept	22.00***	28.91***	25.58***
$D4^{BOX}$			7.93***
EQUT	0.56***	0.53***	0.54***
NI	0.69	2.54***	2.53***
$D4^{BOX}$*EQUT			-
expected sign			positive
$D4^{BOX}$ *NI			-1.88***
expected sign			negative
DY99	-	-	-
DY00	-6.37*	-15.14***	-12.51***
DY01	-	-20.30***	-16.74***
DY02	-12.90***	-26.37***	-23.37***
DY03	-	-21.68***	-17.73***
Adj. R^2	0.332	0.527	0.498
Incremental R^2 (NI given EQUT)	0	0.079	
Incremental R^2 (EQUT given NI)	0.078	0.096	

Test EQUT: [d]

H_0: $b_{1,\text{Equation Xa}} = b_{1,\text{Equation Xb}}$ t-statistic : 0.26

Test NI: [d]

H_0: $b_{2,\text{Equation Xa}} = b_{2,\text{Equation Xb}}$ t-statistic : -2.97***

[a] Coefficients are reported for the variables; * (**, ***): significant at the two-tailed 10% (5%, 1%) level. Standard errors are heteroscedasticity-corrected (White 1980); t-values are

standard normally distributed.

[b] Only significant dummy variables are reported. Insignificant ones are dropped.

[c] Variables are as defined in section 5.2.1 and 5.2.3. $D4^{BOX}$ =1 if company has a high earnings adjusted creditworthiness, 0 otherwise.

[d] Assumption: $\text{cov}\left(b_{\text{Equation A}}; b_{\text{Equation B}}\right) = 0$

The dummy variable $D4^{BOX}$ and the new cut-off rate are used to test Hypothesis 5.2 again. The results from the group comparisons and the model including dummy variables are reported in the table below. Obviously, the results are highly similar to the results presented in Table 17. This leads to the conclusion that the creditworthiness adjustment that is performed in section 5.2.3.2 is robust.

List of Figures

List of Tables

References

Abarbanell, J. (1991), Do Analysts' Earnings Forecasts Incorporate Information in Prior Stock Price Changes?, *Journal of Accounting and Economics*, 147-165.

Abarbanell, J. and V. Bernard (2001), Is the US Stock Market Myopic?, *Journal of Accounting Research*, 221-242.

Achleitner, A. (2004), Venture Capital/Private Equity-Studie 2004: Company (E)valuation und EVCA Valuation Guidelines, *Finanz-Betrieb*, 701-709.

Adrian, F. (2005), Price-Earnings-to-Growth-Ratio, in: Krolle, S., G. Schmitt, and B. Schwetzler (Eds.), *Multiplikatorverfahren in der Unternehmensbewertung*, Stuttgart, 79-86.

Albrecht, T. (2004), Überlegungen zu Endwertermittlung und Wachstumsabschlag, *Finanz-Betrieb*, 732-741.

Aldrich, J. and F. Nelson (1984), *Linear Probability, Logit, and Probit Models*, Beverly Hills.

Alford, A. (1992), The Effect of the Set of Comparable Firms on the Accuracy of the Price-Earnings Valuation Method, *Journal of Accounting Research*, 94-108.

Almus, M. and D. Czarnitzki (2001), *The Effects of Public F&E Subsidies on Firms' Innovation Activities: The Case of Eastern Germany*, ZEW Discussion Paper 01-10, Mannheim.

American Society of Appraisers (ASA) (2002), *Business Valuation Standards*, Herndon, Virginia.

Amran, M. and N. Kulatilaka (1999), *Real Options – Managing Strategic Investment in an Uncertain World*, Boston.

Ang, A. and J. Liu (2001), A General Affine Earnings Valuation Model, *Review of Accounting Studies*, 397-425.

Angrist, J. (1998), Estimating the Labour Market Impact of Voluntary Military Service Using Social Security Data, *Econometrica*, 249-288.

Arbel, A. and P. Strebel (1983), Pay Attention to Neglected Firms!, *Journal of Portfolio Management*, 37-42.

Ariel, R. (1998), Risk Adjusted Discount Rates and the Present Value of Risky Costs, *The Financial Review*, 17-30.

Arnold, J. and K. Hussinger (2004), *Export Behavior and Firm Productivity in German Manufacturing*, ZEW Discussion Paper 04-12, Mannheim.

Arrow, K. (1964), The Role of Securities in the Optimal Allocation of Risk Bearing, *Review of Economic Studies*, 91-96.

Association for Investment Management and Research (AIMR) (1999), *Standards of Practice Handbook*, 8th edition, Charlottesville.

Athanassakos, G. and M. Robinson (1994), The Day-of-the-Week Anomaly: The Toronto Stock Exchange Experience, *Journal of Business Finance and Accounting*, 833-856.

Baetge, J., K. Niemeyer, and J. Kümmel (2005), Darstellung der Discounted-Cashflow-Verfahren (DCF-Verfahren) mit Beispiel, in: Peemöller, V. (Ed.), *Praxishandbuch der Unternehmensbewertung*, 3rd edition, Herne, 265-362.

Baker, M. and S. Ruback (1999), *Estimating Industry Multiples*, Working Paper, Harvard University.

Ball, R. and P. Brown (1968), An Empirical Evaluation of Accounting Income Numbers, *Journal of Accounting Research*, 159-178.

Ball, R., S. Kothari, and C. Wasley (1995), Can We Implement Research on Stock Trading Rules, *Journal of Portfolio Management*, 54-63.

Ballwieser, W. (1991), Unternehmensbewertung mit Hilfe von Multiplikatoren, in: Rückle, D. (Ed.), *Aktuelle Fragen der Finanzwirtschaft und der Unternehmensbesteuerung. Festschrift zum 70. Geburtstag von Erich Loitlsberger*, Vienna, 47-66.

Ballwieser, W. (1997), Eine neue Lehre der Unternehmensbewertung?, *Der Betrieb*, 185-191.

Bamber, L., T. Christensen, and K. Gaver (2000), Do We Really "Know" What We Think We Know? A Case Study of Seminal Research and Its Subsequent Overgeneralization, *Accounting Organization and Society*, 103-129.

Bamberger, B. (1999), Unternehmensbewertung in Deutschland: Die zehn häufigsten Bewertungsfehler, *Betriebswirtschaftliche Forschung und Praxis*, 653-670.

Banz, R. (1981), The Relationship Between Return and Market Value of Common Stocks, *Journal of Financial Economics*, 3-18.

Barberis, N. and R. Thaler (2002), *A Survey of Behavioural Finance*, Working Paper, National Bureau of Economic Research, Cambridge.

Barth, M., W. Beaver, and W. Landsman (1998), Relative Valuation Roles of Equity Book Value and Net Income as a Function of Financial Health, *Journal of Accounting and Economics*, 1-34.

Barth, M., J. Elliot, and M. Finn (1999), Market Rewards Associated with Patterns of Increasing Earnings, *Journal of Accounting Research*, 387-413.

Barthel, C. (1996), Unternehmenswert: Die Vergleichsorientierten Bewertungsverfahren, *Der Betrieb*, 149-163.

Barthel, C. (2002), Handbuch der Unternehmensbewertung – Praxis und Theorie, Auszug aus dem Loseblattwerk Barthel, *Handbuch der Unternehmensbewertung*, Karlsfeld.

Bartov, E., S. Goldberg, and M. Kim (2001), The Valuation-Relevance of Earnings and Cash Lows: An International Perspective, *Journal of International Financial Management and Accounting*, 103- 132.

Basu, S. (1977), Investment Performance and Common Stocks in Relation to Their Price-Earnings Ratios: A Test of the Efficient Market Hypothesis, *Journal of Finance*, 663-682.

Basu, S. (1997), The Conservatism Principle and the Asymmetric Timeliness of Earnings, *Journal of Accounting and Economics*, 3-37.

Bausch, A. (2000), Die Multiplikator-Methode, *Finanz-Betrieb*, 448-459.

Beard, C. and R. Sias (1997), Is There a Neglected Firm Effect?, *Financial Analysts Journal*, 19-23.

Beatty, R., S. Riffe, and R. Thompson (1999), The Method of Comparables and Tax Court Valuations of Private Firms: An Empirical Investigation, *Accounting Horizons*, 177-199.

Beaver, W. (1968), The Information Content of Annual Earnings Announcements, *Journal of Accounting Research*, 67-92.

Beaver, W. (1999), Comments on "An Empirical Assessment of the Residual Income Valuation Model", *Journal of Accounting and Economics*, 35-42.

Beaver, W., R. Lambert, and D. Morse (1980), The Information Content of Security Prices, a Second Look, *Journal of Accounting and Economics*, 139-157.

Beaver, W. and D. Morse (1978), What Determines Price-Earnings Ratios?, *Financial Analysts Journal*, 65-78.

Beckmann, C., J. Meister, and M. Meitner (2003), Das Multiplikatorverfahren in der kapitalmarktorientierten Unternehmensbewertungspraxis, *Finanz-Betrieb*, 103-105.

Behringer, S. (2002), *Unternehmensbewertung der Mittel- und Kleinbetriebe*, 2nd edition, Berlin.

Beiker, H. (1993), *Überrenditen und Risiken kleiner Aktiengesellschaften*, Cologne.

Benninga, S. and O. Sarig (1997), *Corporate Finance: A Valuation Approach*, New York.

Berck, P. and K. Sydsæter (2000), *Economists' Mathematical Manual*, 3rd edition, Berlin.

Berger, P., E. Ofek, and I. Swary (1996), Investor Valuation of the Abandonment Option, *Journal of Financial Economics*, 257-287.

Berk, J., R. Green, and V. Naik (1999), Optimal Investment, Growth Options and Security Returns, *Journal of Finance*, 1553-1607.

Bernard, V. (1989), Capital Market Research in Accounting During the 1980s: A Credit Review, in: Frecka, T. (Ed.), *The State of Accounting Research as We Enter the 1990's*, University of Illinois, 72-120.

Bernstein, L. and J. Wild (1998), *Financial Statement Analysis: Theory, Application, and Interpretation*, Boston.

Berry, R. and R. Dyson (1980), On the Negative Risk Premium for Risk Adjusted Discount Rates, *Journal of Business Finance and Accounting*, 427-436.

Bessembinder, H. and K. Chan (1998), Market Efficiency and Returns to Technical Analysis, *Financial Management*, 5-17.

Bhagat, S. and D. Hirshleifer (1996), *Do Takeovers Create Value? An Intervention Approach*, Working Paper, University of Michigan Graduate School of Business.

Bhojraj, S. and C. Lee (2002), Who Is My Peer? A Valuation-Based Approach to the Selection of Comparable Firms, *Journal of Accounting Research*, 407-439.

Biddle, G., G. Seow, and A. Siegel (1995), Relative Versus Incremental Information Content, *Contemporary Accounting Research*, 1-23.

Binger, B. and E. Hoffmann (1998), *Microeconomics with Calculus*, 2nd edition, Reading, Massachussets.

Bjerksund, P. (1991), *Contingent Claims Evaluation When the Convenience Yield Is Stochastic: Analytical Results*, Working Paper, Norwegian School of Economics and Business Administration.

Black, E. (1998), *Which Is More Value Relevant: Earnings or Cash Flows? A Life Cycle Examination*, Working Paper, University of Arkansas.

Black, F. (1972), Capital Market Equilibrium with Restricted Borrowing, *Journal of Business*, 444–455.

Black, F. (1986), Noise, *Journal of Finance*, 529-543.

Black, F. and M. Scholes (1973), The Pricing of Options and Corporate Liabilities, *Journal of Political Economy*, 637-659.

Blanchard, O. (1997), *Macroeconomics*, Upper Saddle River, New Jersey.

Böcking, H. (2003), Zur Bedeutung des Börsenkurses für die angemessene Barabfindung, in: Richter, F. (Ed.), *Kapitalgeberansprüche, Marktwertorientierung und Unterneh-*

menswert: Festschrift für Prof. Dr. Dr. h.c. Jochen Drukarczyk zum 65. Geburtstag, Munich, 59-91.

Böcking, H. and K. Nowak (1999), Marktorientierte Unternehmensbewertung, *Finanz-Betrieb*, 169-176.

Bonbright, J. (1965), *Valuation of Property*, Vol. 1, Charlottesville.

Booth, L. (1982), Correct Procedures for the Evaluation of Risky Cash Outflows, *Journal of Financial and Quantitative Analysis*, 287-300.

Booth, R. (2001), *Minority Discounts and Control Premiums in Appraisal Proceedings*, Working Paper, University of Maryland, Baltimore.

Born, K. (1995), *Unternehmensanalyse und Unternehmensbewertung*, Stuttgart.

Box, G. and D. Cox (1964), An Analysis of Transformations, *Journal of the Royal Statistical Society*, 211-243.

Bräutigam, C. (2004), *Effizienzprobleme an Finanzmärkten*, Marburg.

Brav, A. and J. Heaton (2003), *Market Indeterminacy*, Working Paper, Duke University.

Brealy, R. and S. Myers (2003), *Principles of Corporate Finance*, 7th edition, New York.

Brennan, M. (1970), Taxes, Market Valuation and Corporate Financial Policy, *National Tax Journal*, 417-427.

Brief, R. and P. Zarowin (1999), *The Value Relevance of Dividends, Book Value and Earnings*, Working Paper, New York University Department of Accounting.

Brown, L. (1996), Analyst Forecasting Errors and Their Implications for Security Analysis: An Alternative Perspective, *Financial Analysts Journal*, 40-47.

Brown, L. (1997), Analyst Forecasting Errors: Additional Evidence, *Financial Analysts Journal*, 81-88.

Brown, L. (1998), *Managerial Behavior and the Bias in Analysts' Forecasts*, Working Paper, Georgia State University.

Bruns, C. (1994), *Bubbles und Excess Volatility auf dem Deutschen Aktienmarkt*, Wiesbaden.

Brush, J. (1986), Eight Relative Strength Models Compared, *Journal of Portfolio Management*, 21-28.

Buchner, R. and J. Englert (1994), Die Bewertung von Unternehmen auf der Basis des Unternehmensvergleichs, *Der Betriebs-Berater*, 1573-1580.

Burgstahler, D. and I. Dichev (1997), Earnings, Adaptation and Equity Value, *The Accounting Review*, 187-215.

Burmeister, E., R. Roll, and S. Ross (1994), A Practitioner's Guide to Arbitrage Pricing Theory, *Finanzmarkt und Portfolio Management*, 312-331.

Bushee, B. (2001), Do Institutional Investors Prefer Near-Term Earnings over Long-Run Value?, *Contemporary Accounting Research*, 207-246.

Busse von Colbe, W. (1957), *Der Zukunftserfolg*, Wiesbaden.

Busse von Colbe, W. (2000), *Ergebnis je Aktie nach DVFA-SG*, 3rd edition, Stuttgart.

Campbell, J., A. Lo, and A. MacKinlay (1997), *The Econometrics of Financial Markets*, Princeton.

Campbell, J. and R. Shiller (1988a), Stock Prices, Earnings, and Expected Dividends, *Journal of Finance* 43, 661-676.

Campbell, J. and R. Shiller (1988b), The Dividend-Price Ratio and Expectations of Future Dividends and Discount Factors, *Review of Financial Studies*, 195-227.

Casey, C. (2000), *Unternehmensbewertung und Marktpreisfindung*, Wiesbaden.

Chan, L., J. Karceski, and J. Lakonishok (2003), The Level and Persistence of Growth Rates, *Journal of Finance*, 643-684.

Charitou, A., C. Clubb, and A. Andreou (2000), The Value Relevance of Earnings and Cash Flows: Empirical Evidence for Japan, *Journal of International Financial Management and Accounting*, 1-22.

Charitou, A., C. Clubb, and A. Andreou (2001), The Effect of Earnings Permanence, Growth and Firm Size on the Usefulness of Cash Flows and Earnings in Explaining Security Returns: Empirical Evidence for the UK, *Journal of Business Finance and Accounting*, 563-594.

Chen, N. (1991), Financial Investment Opportunities and the Macroeconomy, *Journal of Finance*, 529-594.

Chen, P. and G. Zhang (2002), *The Roles of Earnings and Book Value in Equity Valuation: A Real Option Based Analysis*, Working Paper, Hong Kong University.

Cheng, A. and R. McNamara (2000), The Valuation Accuracy of the Price-Earnings and Price-Book Benchmark Valuation Methods, *Review of Quantitative Finance and Accounting*, 349-370.

Cheng, C., C. Liu, and T. Schaefer (1996), Earnings Permanence and the Incremental Information Content of Cash Flows from Operations, *Journal of Accounting Research*, 173-181.

Cheng, C. and T. Liu (1997), The Value-Relevance of SFAS No. 95 Cash Flows from Operations as Assessed by Security Market Effects, *Accounting Horizons*, 1-15.

Cheng, Q. (2004), *The Role of Analysts' Forecasts in Accounting-Based Valuation: A Critical Evaluation*, Working Paper, The University of British Columbia.

Cheredito, Y. and T. Hadewicz (2001), Marktorientierte Unternehmensbewertung, *Der Schweizer Treuhänder*, 321-330.

Choudhary, P. (2004), *The Influence of Across-Firm Differences in Accounting Quality on Investors' Choice of Multiples*, Working Paper, Duke University.

Christie, A. (1987), On Cross-Sectional Analysis in Accounting Research, *Journal of Accounting and Economics*, 231-258.

Cochran, W. and D. Rubin (1973), Controlling Bias in Observational Studies: A Review, *Sankhya*, 417-446.

Cochrane, J. (1991), Volatility Tests and Efficient Markets, *Journal of Monetary Economics*, 463-485.

Coenenberg, A. and W. Schultze (2002), Das Multiplikator-Verfahren in der Unternehmensbewertung: Konzeption und Kritik, *Finanz-Betrieb*, 697-703.

Collins, D. and S. Kothari (1989) An Analysis of Intertemporal and Cross-Sectional Determinants of Earnings Response Coefficients, *Journal of Accounting and Economics*, 143-181.

Collins, D., E. Maydew, and I. Weiss (1997), Changes in the Value Relevance of Earnings and Book Values over the Past Forty Years, *Journal of Accounting and Economics*, 39-67.

Collins, D. M. Pincus, and H. Xie (1999), Equity Valuation and Negative Earnings: The Role of Book Value of Equity, *The Accounting Review*, 29-61.

Conrad, J. and G. Kaul (1988), Time Variation in Expected Returns, *Journal of Business*, 409-425.

Conroy, R., K. Eades, and T. Harris (2000), A Test of the Relative Pricing Effects of Dividends and Earnings: Evidence from Simultaneous Announcements in Japan, *The Journal of Finance*, 1199-1227.

Cook, R. and S. Weisberg (1983), Diagnostics for Heteroscedasticity in Regression, *Biometrika*, 1-10.

Copeland, T., J. Weston, and K. Shastri (2005), *Financial Theory and Corporate Policy*, 4th edition, Reading.

Copeland, T., T. Koller, and J. Murrin (2000), *Valuation, Measuring and Managing the Value of Companies*, 3rd edition, New York.

Courteau, L., J. Kao, and G. Richardson (2001), Equity Valuation Employing The Ideal Versus Ad Hoc Terminal Value Expressions, *Contemporary Accounting Research*, 625-661.

Creutzmann, A. and N. Deser (2005), Einführung, in: Krolle, S., G. Schmitt, and B. Schwetzler (Eds.), *Multiplikatorverfahren in der Unternehmensbewertung*, Stuttgart, 1-4.

Czarnitzki, D., B. Ebersberger, and A. Fier (2004), *The Relationship Between R&D Collaboration, Subsidies and Patenting Activity: Empirical Evidence from Finland and Germany*, ZEW Discussion Paper 04-37, Mannheim.

Czarnitzki, D. and K. Kraft (2004), *Are Credit Ratings Valuable Information?*, ZEW Discussion Paper 04-07, Mannheim.

Damodaran, A. (1996), *Investment Valuation*, New York.

Damodaran, A. (1997), *Corporate Finance, Theory and Practice*, New York.

Damodaran, A. (2001a), *The Dark Side of Valuation*, New York.

Damodaran, A. (2001b), *Corporate Finance, Theory and Practice*, 2nd edition, New York.

Damodaran, A. (2002), *Investment Valuation*, 2nd edition, New York.

De Long, J., A. Shleifer, L. Summers, and R. Waldmann (1990a), Positive Feedback Investment Strategies and Destabilizing Rational Speculation, *Journal of Finance*, 379-395.

De Long, J., A. Shleifer, L. Summers, and R. Waldmann (1990b), Noise Trader Risk in Financial Markets, *Journal of Political Economy*, 703-738.

DeBondt, W. and R. Thaler (1985), Does the Stock Market Overreact?, *Journal of Finance*, 793-805.

Dechow, P. (1994), Accounting Earnings and Cash Flows as Measures of Firm Performance: The Role of Accounting Accruals, *Journal of Accounting and Economics*, 3-42.

Dechow, P., A. Hutton, and R. Sloan (1999), An Empirical Assessment of the Residual Income Valuation Model, *Journal of Accounting and Economics*, 1-34.

DeFusco, R., D. McLeavey, J. Pinto, and D. Runkle (2001), *Quantitative Methods for Investment Analysis*, Baltimore.

Dehejia, R. and S. Wahba (1999), Causal Effects in Nonexperimental Studies: Re-Evaluating the Evaluation of Training Programs, *Journal of the American Statistical Association*, 1053-1062.

Demsetz, H. (1968), The Cost of Transacting, *Quarterly Journal Economics*, 33-53.

Dennis, P., S. Perfect, K. Snow, and K. Wiles (1995), The Effects of Rebalancing on Size and Book-to-Market Ratio Portfolio Returns, *Financial Analysts Journal*, 47-57.

Desai, H. and P. Jain (1995), An Analysis of the Recommendations of the 'Superstars' Money Managers at Barron's Annual Roundtable, *Journal of Finance*, 1257-1273.

Deutscher Investment Trust (DIT) (2004), *Dynamic Investment Trends – Investmentzyklus (II): Sensitivität – vom Beta lernen*, April.

Deutsches Aktieninstitut (DAI) (2003), *Gateway to Euroland – The German Financial Marketplace*, Centre of Euro Area Capital Markets, Frankfurt/M.

Dittmann, I. and C. Weiner (2005), *Selecting Comparables for the Valuation of European Firms*, Working Paper, Humboldt University Berlin.

Dong, M. and D. Hirshleifer (2004), *A Generalized Earnings-Based Stock Valuation Model*, Working Paper, York University.

Dothan, M. (1990), *Prices in Financial Markets*, Oxford.

Drukarczyk, J. (2003), *Unternehmensbewertung*, 4th edition, Munich.

Dubois, M. and P. Louvet (1996), The Day-of-the-Week Effect: the International Evidence, *Journal of Banking and Finance*, 1463-1485.

Duerr, M. (1995), *Investor Relations*, 2nd edition, Munich.

Easton, P., T. Harris, and J. Ohlson (1992), Aggregate Accounting Earnings Can Explain Most of Security Returns: The Case of Long Event Windows, *Journal of Accounting & Economics*, 119-142.

Easton, P., P. Shroff, and G. Taylor (2000), Permanent and Transitory Earnings, Accounting Recording Lag, and the Earnings Coefficient, *Review of Accounting Studies*, 281-300.

Ecker, M. and T. Ehren (2004), Strategische Überlegungen zur Preisgestaltung bei Öffentlichen Übernahmen, in: Lucks, K. (Ed.), *Mergers & Acquistions: Jahrbuch*, 87-90.

Edwards, E. and P. Bell (1961), *The Theory and Measurement of Business Income*, New York.

Eidel, U. (2000), *Moderne Verfahren der Unternehmensbewertung und Performance-Messung*, 2nd edition, Herne, Berlin.

Elton, E., M. Gruber, S. Brown, and W. Goetzmann (2003), *Modern Portfolio Theory and Investment Analysis*, 6th edition, Upper Saddle River, New Jersey.

Engels, W. (1962), *Betriebswirtschaftliche Bewertungslehre im Licht der Entscheidungstheorie*, Cologne.

European Private Equity & Venture Capital Association (EVCA) (2005), *International Private Equity and Venture Capital Valuation Guidelines*, Brussels.

Fama, E. (1965), The Behaviour of Stock Market Prices, *Journal of Business*, 34-105.

Fama, E. (1970), Efficient Capital Markets: A Review of Theory and Empirical Work, *Journal of Finance*, 383-417.

Fama, E. (1976), *Foundations of Finance*, New York.

Fama, E. (1991), Efficient Capital Markets II, *Journal of Finance*, 1575-1617.

Fama, E. (1996), Discounting Under Uncertainty, *Journal of Business*, 415-428.

Fama, E. and K. French (1988), Permanent and Temporary Components of Stock Prices, *Journal of Political Economy*, 246-273.

Fama, E. and K. French (1992), The Cross-Section of Expected Stock Returns, *Journal of Finance*, 427-465.

Fama, E. and K. French (1993), Common Risk Factors in the Returns on Stocks and Bonds, *Journal of Financial Economics*, 3-56.

Fama, E. and K. French (1996), Multifactor Explanation of Asset Pricing Anomalies, *Journal of Finance*, 55-84.

Fama, E. and K. French (2000), Forecasting Profitability and Earnings, *Journal of Business*, 161-175.

Fama, E. and K. French (2003), *The Capital Asset Pricing Model – Theory and Evidence*, CRSP Working Paper, Tuck Business School.

Fama, E. and K. French (1972), *The Theory of Finance*, New York.

Fama, E. and K. French (1998), Value Versus Growth: The International Evidence, *Journal of Finance*, 1975-1999.

Fama, E. and J. McBeth (1973), Risk, Return and Equilibrium: Empirical Tests, *Journal of Political Economy*, 607-636.

Feltham, G. and J. Ohlson (1995), Valuation and Clean Surplus Accounting for Operating and Financial Activities, *Contemporary Accounting Research*, 689-731

Fernandez, P. (2002), Valuation Using Multiples – How Do Analysts Reach Their Conclusions?, Working Paper, IESE, University of Navarra.

Fernandez, P. (2003). *Three Residual Income Valuation Methods and Discounted Cash Flow Valuation*, Working Paper, IESE, University of Navarra.

Financial Accounting Standards Board (FASB) (1978a), *Statement of Financial Accounting Concepts 1 (SFAC 1), Objectives of Financial Reporting by Business Enterprises*, New York.

Financial Accounting Standards Board (FASB) (1978b), *Statement of Financial Accounting Concepts 5 (SFAC 5), Recognition and Measurement in Financial Statements of Enterprises*, New York.

Financial Accounting Standards Board (FASB) (2002), *Statements of Financial Accounting Concepts*, 2002/2003 edition, New York.

Finnerty, J. (1976), Insiders and Market Efficiency, *Journal of Finance*, 1141-1148.

Fortune, P. (1991), Stock Market Efficiency: An Autopsy?, *New England Economic Review*, 17-40.

Foster, G., C. Olsen, and T. Shevlin (1984), Earnings Releases, Anomalies and the Behaviour of Security Returns, *Accounting Review*, 574-603.

Francis, J., P. Olsson, and D. Oswald (2000), Comparing the Accuracy and Explainability of Dividend, Free Cash Flow and Abnormal Earnings Equity Value Estimates, *Journal of Accounting Research*, 45-70.

Francis, J. and K. Schipper (1999), Have Financial Statements Lost Their Relevance?, *Journal of Accounting Research*, 319-352

Franks, J., R. Harris, and C. Mayer (1988), Means of Payment in Takeovers: Results for the United Kingdom and the United States, in: Auerbach, Alan J. (Ed.), *Corporate Takeovers: Causes and Consequences*, Chicago, London.

Frantzmann, J. (1987), Der Montagseffekt am Deutschen Aktienmarkt, *Zeitschrift für Betriebswirtschaft*, 611-635.

French, K. (1980), Stock Returns and the Weekend Effect, *Journal of Financial Economics*, 55-69.

Friedlob, G. and L. Schleifer (2003), *Essentials of Financial Analysis*, New Jersey.

Froot, K. and E. Dabora (1999), How Are Stock Prices Affected by the Location of Trade, *Journal of Financial Economics*, 189-216.

Fuller, R. and C. Hsia (1984), A Simplified Common Stock Valuation Model, *Financial Analysts Journal*, 49-56.

Garz, H. (2000), *Prognostizierbarkeit von Aktienrenditen*, Wiesbaden.

Gaughan, P. (1999), *Mergers, Acquistions and Corporate Restructuring*, 2nd edition, New York.

Gerfin, M. and M. Lechner (2002), A Microeconometric Evaluation of the Active Labour Market Policy in Switzerland, *Economic Journal*, 854-893.

Gerke, W. (1997), Herrschaft der Androiden? Konsequenzen der Kapitalmarkttheorie für das Anlegerverhalten, in: Jünemann, B. (Ed.), *Psychologie für Börsenprofis*, Stuttgart, 19-40.

Gerke, W. (2001), Portfoliotheorie, in: Gerke, W. and M. Steiner (Eds.) *Handwörterbuch des Bank- und Finanzwesens*, 3rd edition, Stuttgart, 1694-1707.

Gerke, W. (2002), Dilemma der Unternehmensbewertung, *Betriebs-Berater*, 1.

Gerke, W., S. Arneth, and J. Fleischer (1999), Kursgewinne bei Aufnahme in den Dax 100, Verluste bei Entnahmen: Indexeffekt am deutschen Kapitalmarkt, *Zeitschrift für Betriebswirtschaft*, 45-66.

Gerke, W. and M. Bank (2003), *Finanzierung*, 2nd edition, Stuttgart.

Gerke, W. and H. Rapp (1994), Strukturveränderungen im internationalen Börsenwesen, *Die Betriebswirtschaft*, 5-23.

Gibson, R. and E. S. Schwartz (1990), Stochastic Convenience Yield and the Pricing of Oil Contingent Claims, *The Journal of Finance*, 959-976.

Gilson, R. and R. Kraakman (2003), *The Mechanisms of Market Efficiency Twenty Years Later, the Hindsight Bias*, Discussion Paper, Harvard Law School.

Gilson, S., E. Hotchkiss, and R. Ruback (2000), Valuation of Bankrupt Firms, *Review of Financial Studies*, 43-47.

Givoly, D. and D. Palmon (1985), Insider Trading and the Exploitation of Inside Information: Some Empirical Evidence, *Journal of Business*, 69-87.

Gordon, M. and E. Shapiro (1956), Capital Equipment Analysis: The Required Rate of Profit, *Management Science*, 102-110.

Graham, J. and C. Harvey (2001) The Theory and Practice of Corporate Finance: Evidence from the Field, *Journal of Financial Economics*, 187-243.

Greene, W. (1997), *Econometric Analysis*, 3rd edition, New Jersey.

Großfeld, B. (2000), Börsenkurs und Unternehmenswert, *Betriebs-Berater*, 261-266.

Grossman, S. and J. Stiglitz (1980), On the Impossibility of Informationally Efficient Markets, *American Economic Review*, 123-136.

Gruber, A. (1988), *Signale, Bubbles und Rationale Anlagestrategien auf Kapitalmärkten*, Wiesbaden.

Gwartney, J. and R. Stroup (1997), *Economics, Private and Public Choice*, 8th edition, Fort Worth.

Hachmeister, D. (2000), *Der Discounted Cash Flow als Maß der Unternehmenswertsteigerung*, 4th edition, Frankfurt/M.

Hagerman, R. and R. Richmond (1973), Random Walks, Martingales and the OTC, *Journal of Finance*, 897-909.

Hail, L. and C. Meyer (2002), Unternehmensbewertung, *Der Schweizer Treuhänder*, 573-584.

Hanouna, P., A. Sarin, and A. Shapiro (2001), *Value of Corporate Control: Some International Evidence*, Working Paper, Marshall School of Business, Los Angeles.

Harris, T., M. Lang, and H. Möller (1994), The Value Relevance of German Accounting Measures: An Empirical Analysis, *Journal of Accounting Research*, 187-209.

Heckman, J., H. Ichimura, and P. Todd (1998), Matching as an Econometric Evaluation Estimator, *Review of Economic Studies*, 261-294.

Heckman, J., R. Lalonde, and J. Smith (1999), The Economics and Econometrics of Active Labour Market Programs, in: Ashenfelter, A. and D. Card (Eds.), *Handbook of Labour Economics*, Amsterdam, 1866-2097.

Heller, M. (2001), M&A: Globale Tendenz setzt sich fort, *Börsen-Zeitung*, January 27, B5.

Henselmann, K. (2000), Der Restwert in der Unternehmensbewertung – eine „Kleinigkeit"?, *Finanz-Betrieb*, 151-157.

Hering, T. (1999), *Finanzwirtschaftliche Unternehmensbewertung*, Wiesbaden.

Hering, T. (2000), Konzeption der Unternehmensbewertung und ihre Eignung für mittelständische Unternehmen, *Betriebswirtschaftliche Forschung und Praxis*, 433-453.

Hering, T. (2003), *Investitionstheorie*, 2nd edition, Munich.

Herrmann, V. (2002), *Marktpreisschätzungen mit kontrollierten Multiplikatoren*, Lohmar.

Hickmann, K. and G. Petry (1990), A Comparison of Stock Price Predictions Using Court Accepted Formulas, Dividend Discount and P/E Models, *Financial Management*, 76-87.

Hietala, P., S. Kaplan, and T. Robinson (2000), *What Is The Price of Hubris? Using Take-over Battles to Infer Overpayments and Synergies*, Working Paper, Insead and University of Chicago.

Hillebrandt, F. (2001), Multiplikatorverfahren, *Die Betriebswirtschaft*, 618 - 621.

Hirshleifer, J. (1958), On the Theory of Optimal Investment Decision, *The Journal of Political Economy*, 329-352.

Hirshleifer, J. and D. Hirshleifer (1997), *Price Theory and Applications*, 6th edition, New Jersey.

Hodrick, R. and E. Prescott (1997), Postwar U.S. Business Cycles: An Empirical Investigation, *Journal of Money, Credit and Banking*, 1-16.

Hogg, R. and A. Craig (1995), *Introduction to Mathematical Statistics*, New Jersey.

Holthausen, R. and R. Watts (2001): The Relevance of the Value-Relevance Literature for Financial Accounting Standard Setting, *Journal of Accounting and Economics*, 3-75

Holzer, E. (2001), *Informationseffizienz des deutschen und österreichischen Kapitalmarktes*, Vienna.

Hommel, M. and I. Braun (2002), Marktorientierte Unternehmensbewertung – Börsenkurs auf dem Prüfstand, *Betriebs-Berater*, 10-17.

Hommel, U. and J. Mueller (1999), Realoptionsbasierte Investitionsbewertung, *Finanz-Betrieb*, 177-188.

Hostettler, S. (2000), *Economic Value Added*, 4th edition, Vienna.

Hull, J. (2003), *Options, Futures, and Other Derivatives*, 5th edition, New Jersey.

Hung, M. and K. Subramanyam (2004), *Financial Statement Effects of Adopting International Accounting Standards: The Case of Germany*, Working Paper, University of Southern California.

Institut der Wirtschaftsprüfer (IDW) (1998), *Wirtschaftsprüfer-Handbuch*, Vol. 2, 11th edition, Düsseldorf.

Institut der Wirtschaftsprüfer (IDW) (2000), IDW Standard: Grundsätze zur Durchführung von Unternehmensbewertungen (IDW S 1), *Die Wirtschaftsprüfung*, 825-842.

Institut der Wirtschaftsprüfer (IDW) (2002), *Wirtschaftsprüfer-Handbuch*, Vol. 2, 12th edition, Düsseldorf.

Institut der Wirtschaftsprüfer (IDW) (2005), IDW ES 1 n.F. – Grundsätze zur Durchführung von Unternehmensbewertungen, *Die Wirtschaftsprüfung*, 28-46.

Jaensch, G. (1966), *Wert und Preis der ganzen Unternehmung*, Cologne.

James, C. and R. Edmister (1983), The Relation Between Common Stock Returns, Trading Activity, and Market Value, *Journal of Finance*, 1075-1086.

Jan, C. and J. Ou (1995), *The Role of Negative Earnings in the Valuation of Equity Stocks*, Working Paper, Santa Clara University.

Jennergren, P. and P. Korsvold (1975), The Non-Random Character of Norwegian and Swedish Stock Market Prices, in: Elton, E. and M. Gruber (Eds.), *International Capital Markets*, Amsterdam.

Jensen, M. (1968), The Performance of Mutual Funds in the Period 1945-1964, *Journal of Finance*, 389-416.

Jensen, M. (1978), Some Anomalous Evidence Regarding Market Efficiency, *Journal of Financial Economics*, 95-101.

Jonas, M., A. Löffler, and J. Wiese (2004), Das CAPM mit Deutscher Einkommensteuer, *Die Wirtschaftsprüfung*, 898-906.

Jones, C. (1998), *Investments*, 6th edition, New York.

Kames, C. (2000), *Unternehmensbewertung durch Finanzanalysten als Ausgangspunkt eines Value Based Measurement*, Frankfurt/M.

Kaplan, S. and R. Ruback (1995), The Valuation of Cash Flow Forecasts: An Empirical Analysis, *Journal of Finance*, 1059-1093.

Keim, D. and R. Stambaugh (1984), A Further Investigation of Weekend Effects in Stock Returns, *Journal of Finance*, 819-840.

Keim, D. and R. Stambaugh (1986), Predicting Returns in Stock and Bond Markets, *Journal of Financial Economics*, 357-390.

Kim, M. and J. Ritter (1999), Valuing IPOs, *Journal of Financial Economics*, 409–437.

Kim, O. and R. Verrecchia (1991), Trading Volume and Price Reaction to Public Announcements, *Journal of Accounting Research*, 302-321.

Kothari, S., J. Shanken, and R. Sloan (1995), Another Look at the Cross Section of Expected Stock Returns, *Journal of Finance*, 185-224.

Kothari, S. and J. Zimmermann (1995), Price and Return Models, *Journal of Accounting and Economics*, 155-192.

Kothari, S.P. (2001), Capital Markets Research in Accounting, *Journal of Accounting and Economics*, 105-231.

Krolle, S. (2005), Enterprise Value/EBIT(DA)-Verhältnis, in: Krolle, S., G. Schmitt, and B. Schwetzler (Eds.), *Multiplikatorverfahren in der Unternehmensbewertung*, Stuttgart, 21-48.

Krolle, S., G. Schmitt, and B. Schwetzler (2005), *Multiplikatorverfahren in der Unternehmensbewertung*, Stuttgart.

Kruschwitz, L. and A. Löffler (1999), *Sichere und unsichere Steuervorteile bei der Unternehmensbewertung I. – Kritische Anmerkungen (nicht nur) zum WP-Handbuch 1998*, Working Paper, Berlin.

Kuelpmann, M. (2004), *Irrational Exuberance Reconsidered – The Cross Section of Stock Returns*, 2nd edition, Berlin et al.

Kuhlmann, H. (2005), Kurs/Buchwert-Verhältnis, in: Krolle, S., G. Schmitt, and B. Schwetzler (Eds.), *Multiplikatorverfahren in der Unternehmensbewertung*, Stuttgart, 95-102.

Künnemann, M. (1985), *Objektivierte Unternehmensbewertung*, Frankfurt/M.

Kusterer, G. (2003), *Unternehmensbewertungsverfahren zur Emissionspreisbestimmung bei Börsengängen am Neuen Markt*, Frankfurt/M.

Küting, K. and U. Eidel (1999), Marktwertansatz contra Ertragswert- und Discounted Cash-Flow-Verfahren, *Finanz-Betrieb*, 225-231.

Kyle, A.S., 1989, Informed Speculation with Imperfect Competition, *Review of Economic Studies*, 317-356.

La Porta, R. (1996), Expectations and the Cross-Section of Stock Returns, *Journal of Finance*, 1715-1742.

Lakonishok, J. and T. Vermaelen (1990), Anomalous Price Behaviour around Repurchase Tender Offers, *Journal of Finance*, 455-477.

Lamont, O. and R. Thaler (2003), Can the Market Add and Subtract? Mispricing in Tech Stock Carve-Outs, *Journal of Political Economy*, 227-268.

Lander, D. and G. Pinches (1998), Challenges to the Practical Implementation of Modeling and Valuing Real Options, *The Quarterly Review of Economics and Finance*, 537-567.

Lang, M. and M. McNichols (1997), *Institutional Trading and Corporate Performance*, Working Paper, Stanford University.

LeBaron, D., R. Vaitilingham, and M. Pitchford (1999), *The Ultimate Book of Investment Quotations*, Oxford.

Lechner, M. (1999a), Earnings and Employment Effects of Continuous Off-the-Job Training in East Germany After Reunification, *Journal of Business and Economics Statistics*, 74-90.

Lechner, M. (1999b), *Identification and Estimation of Causal Effects of Multiple Treatments Under the Conditional Independence Assumption*, IZA Discussion Paper, Bonn.

Lechner, M. (2000), An Evaluation of Public Sector Sponsored Continuous Vocational Training in East Germany, *Journal of Human Ressources*, 347-375

Lee, C. (1999), Accounting-Based Valuation: A Commentary, *Accounting Horizons*, 413-425.

Lee, W. and M. Solt (1986), Insider Trading: A Poor Guide to Market Timing, *Journal of Portfolio Management*, 65-71.

Leuner, R. (2002), Bewertung ertragsschwacher Unternehmen (Sanierung), in: Peemöller, V. (Ed.), *Praxishandbuch der Unternehmensbewertung*, Herne, 649-670.

Leuven, E. and B. Sianesi (2003), PSMATCH2: *Stata Module to Perform Full Mahalanobis and Propensity Score Matching, Common Support Graphing, and Covariate Imbalance Testing*, Version 1.2.0, http://ideas.repec.org/c/boc/bocode/s432001.html.

Lev, B. (1969) Industry Averages as Targets for Financial Ratios, *Journal of Accounting Research*, 290-299.

Lev, B. and P. Zarowin (1999) The Boundaries of Financial Reporting and How to Extend Them, *Journal of Accounting Research*, 353-385.

Lintner, J. (1965), The Valuation of Risky Assets and the Selection of Risky Investments in Stock Portfolios and Capital Budgets, *Review of Economics and Statistics*, 13-37.

Liu, J., D. Nissim, and J. Thomas (2002), Equity Valuation Using Multiples, *Journal of Accounting Research*, 135-172.

Lo, A. and C. MacKinlay (1988), Stock Market Prices Do Not Follow a Random Walk: Evidence from a Simple Specification Test, *Review of Financial Studies*, 41-46.

Lo, K. and T. Lys (2000a), The Ohlson Model: Contribution to Valuation Theory, Limitations, and Empirical Applications, *Journal of Accounting, Auditing & Finance*, 337-367.

Lo, K. and T. Lys (2000b), *Bridging the Gap Between Value Relevance and Information Content*, Working Paper, University of British Colombia.

Löhnert, P. and U. Böckmann (2005), Multiplikatorverfahren in der Unternehmensbewertung, in: Peemöller, V. (Ed.), *Praxishandbuch der Unternehmensbewertung*, 3rd edition, Herne, 403-428.

Loistl, O. (1994), *Kapitalmarkttheorie*, 3rd edition, Munich.

Luce, R. and H. Raiffa (1957), *Games and Decisions*, New York.

Lundholm, R. (1995), A Tutorial on the Ohlson and Feltham/Ohlson Models: Answers to Some Frequently Asked Questions, *Contemporary Accounting Research*, 749-761.

Maddala, G. (1983), *Limited-Dependent and Qualitative Variables in Econometrics*, Cambridge.

Madden, B. (1999), *CFROI Valuation, A Total System Approach To Valuing the Firm*, Oxford.

Mandl, G. and K. Rabel (1997), *Unternehmensbewertung*, Vienna.

Mandl, G. and K. Rabel (2005), Methoden der Unternehmensbewertung (Überblick), in: Peemöller, V. (Ed.), *Praxishandbuch der Unternehmensbewertung*, 3rd edition, Herne, 47-88.

Matschke, M. (1976), *Der Arbitriumwert der Unternehmung*, Cologne.

McCarter, M. and K. Aschwald (1992), The Market Comparable Approach Using Public Market Data, in: West, T. and J. Jones (Eds.), *Handbook of Business Valuation*, New York, 145-166.

McConnell, J. and C. Muscarella (1985), Corporate Capital Expenditure Decisions and the Market Value of the Firm, *Journal of Financial Economics*, 399-422.

McConnell, J. and S. Wahal (1997), *Do Institutional Investors Exacerbate Managerial Myopia?*, Working Paper, Purdue University and Emory University.

McDonald, R. and D. Siegel (2001), The Value of Waiting to Invest, in: Schwartz, E. and L. Trigeorgis (Eds.), *Real Options and Investment Under Uncertainty: Classical Readings and Recent Contributions*, Cambridge et al., 253-272.

Meitner, M. (2003), Ein Jahr „Deutscher Corporate Governance Kodex" aus der Sicht von Finanzanalysten und institutionellen Investoren, *Finanz-Betrieb*, 763-766.

Meitner, M. (2003a), *Option-Style Multi-Factor Comparable Company Valuation for Practical Use*, ZEW Discussion Paper 03-76, Mannheim.

Meitner, M. (2004), *How Fundamentals Drive the Equity Value*, PricewaterhouseCoopers, Global Competency Center – Research Excellence Series, Discussion Paper, Amsterdam.

Meitner, M. (2005), Die Berücksichtigung und Bestimmung langfristiger Zahlungsüberschusswachstumsraten in der Unternehmensbewertung, *Unternehmensbewertung & Management*, 5-10.

Meitner, M. and P. Westerheide (2005), *The Impact of a Stock Market Downturn on Corporate Financing Activities in Germany*, ZEW Discussion Paper 05-91, Mannheim.

Mellerowicz, K. (1952), *Der Wert des Unternehmens als Ganzes*, Essen.

Menkhoff, L. and C. Roeckemann (1994), Noise Trading auf Aktienmärkten. Ein Überblick zu verhaltensorientierten Erklärungsansätzen nicht-fundamentaler Kursbildung, *Zeitschrift für Betriebswirtschaft*, 277-295.

Merton, R. (1973), An Intertemporal Capital Asset Pricing Model, *Econometrica*, 867-887.

Merton, R. (1987), A Simple Model of Capital Market Equilibrium with Incomplete Information, *Journal of Finance*, 483-510.

Meyer, B. (1994), *Der Overreaction-Effekt am Deutschen Aktienmarkt*, Frankfurt/M.

Milgrom, P. and N. Stokey (1982), Information, Trade, and Common Knowledge, *Journal of Econometric Theory*, 17-27.

Mitchell, M. and E. Stafford (1997), *Managerial Decision and Long-Term Stock Price Performance*, Working Paper, University of Chicago.

Möller, H. (1985), Die Informationseffizienz des deutschen Aktienmarktes – eine Zusammenfassung und Analyse empirischer Untersuchungen, *Zeitschrift für betriebswirtschaftliche Forschung*, 500-518.

Möller, H. and B. Hüfner (2002), Zur Bedeutung der Rechnungslegung für den Deutschen Aktienmarkt – Begründung, Messprobleme und Erkenntnisse empirischer Forschung, *Österreichisches Jahrbuch für Controlling und Rechnungswesen*, 405-463.

Möller, H. and F. Schmidt (1998), Zur Bedeutung von Jahresabschlüssen und DVFA/SG-Daten für die fundamentale Aktienbewertung, in: Moeller, H. and F. Schmidt (Eds.), *Rechnungswesen als Instrument für Führungsentscheidungen, Festschrift für Adolf Coenenberg*, Stuttgart, 477-504.

Moser, U. and S. Auge-Dickhut (2003), Der Informationsgehalt von Marktpreisabschätzungen auf Basis von Vergleichsverfahren, *Finanz-Betrieb*, 10-22.

Mossin, J. (1966), Equilibrium in a Capital Asset Market, *Econometrica*, 768-783.

Moxter, A. (1983), *Grundsätze ordnungsmäßiger Unternehmensbewertung*, 2nd edition, Wiesbaden.

Mueller, D. (1990), *The Dynamics of Company Profits, An International Comparison*, Berlin.

Muenstermann, H. (1970), *Wert und Bewertung von Unternehmen*, 3rd edition, Wiesbaden.

Myers, J. (1999), Implementing Residual Income Valuation with Linear Information Dynamics, *Accounting Review*, 1-28.

Myers, S. (1977), Determinants of Corporate Borrowing, *Journal of Financial Economics*, 147-175.

Myers, S. and S. Majd (1990), Abandonment Value and Project Life, *Advances in Futures and Options Research*, 1-21.

Nelles, M., J. Rojahn, and C. Berner (2001), Unternehmensbewertungsverfahren im Rahmen von Börsengängen am Neuen Markt, *Finanz-Betrieb*, 322-327.

Neumann, J. von and O. Morgenstern (1944), *The Theory of Games and Economic Behaviour*, Princeton.

Nowak, K. (2000), *Marktorientierte Unternehmensbewertung*, Wiesbaden.

O'Byrne, S. (1997), EVA® and Shareholder Return, *Financial Practice and Education*, 50-54.

O'Hara, M. (1997), *Market Microstructure Theory*, Cambridge.

Oehler, A. (1995), Informationseffiziente Finanzmärkte? Empirische Befunde zur Nutzung vergangener Preisveränderungen auf Aktienmärkten, in: Hanusch, H. and W. Gick (Eds.), *Ansätze für ein Neues Denken in der Wirtschaftspolitik*, Munich, 111-135.

Ohlson, J. (1991), The Theory of Value and Earnings, and an Introduction to the Ball-Brown Analysis, *Contemporary Accounting Research*, 1-19.

Ohlson, J. (1995), Earnings, Book Values, and Dividends in Equity Valuation, *Contemporary Accounting Research*, 661-687.

Ohlson, J. (2001), Earnings, Book Values, and Dividends in Equity Valuation: An Empirical Perspective, *Contemporary Accounting Research*, 107-120.

Olbrich, M. (2000), Zur Bedeutung des Börsenkurses für die Bewertung von Unternehmen und Unternehmensanteilen, *Betriebswirtschaftliche Forschung und Praxis*, 454-464.

Ota, K. (2001), *The Value-Relevance of Book Value, Current Earnings, and Management Forecasts of Earnings*, Working Paper, Australian National University.

Ou, J. and S. Penman (1989), Financial Statement Analysis and the Prediction of Stock Returns, *Journal of Accounting & Economics*, 295-329.

Palepu, K., V. Bernard, and P. Healy (1996), *Business Analysis and Valuation: Using Financial Statements*, Cincinnati.

Peasnell, K. (1982), Some Formal Connections Between Economic Values and Yields and Accounting Numbers, *Journal of Business Finance & Accounting*, 361-381.

Peavy, J. and D. Goodman (1983), The Significance of P/Es for Portfolio Returns, *Journal of Portfolio Management*, 43-47.

Peemöller, V. (2003), *Bilanzanalyse und Bilanzpolitik*, 3rd edition, Wiesbaden.

Peemöller, V. (2005a), Wert und Werttheorien, in: Peemöller, V. (Ed.), *Praxishandbuch der Unternehmensbewertung*, 3rd edition, Herne, 1-14.

Peemöller, V. (2005b), Anlässe der Unternehmensbewertung, in: Peemöller, V. (Ed.), *Praxishandbuch der Unternehmensbewertung*, 3rd edition, Herne, 15-25.

Peemöller, V. and C. Beckmann (2005), Der Realoptionsansatz, in: Peemöller, V. (Ed.), *Praxishandbuch der Unternehmensbewertung*, 3rd edition, Herne, 797-811.

Peemöller, V., C. Beckmann, and M. Meitner (2005), Einsatz eines Nachsteuer-CAPM bei der Bestimmung objektivierter Unternehmenswerte – Eine kritische Analyse des IDW ES 1 n.F., *Betriebs-Berater*, 90-96.

Peemöller, V., P. Bömelburg, and A. Denkmann (1994), Unternehmensbewertung in Deutschland: Eine empirische Erhebung, *Die Wirtschaftsprüfung*, 741-749.

Peemöller, V. and S. Kunowski (2005), Ertragswertverfahren nach IDW, in: Peemöller, V. (Ed.), *Praxishandbuch der Unternehmensbewertung*, 3rd edition, Herne, 201-263.

Peemöller, V., S. Kunowski, and J. Hillers (1999), Ermittlung des Kapitalisierungszinses für internationale Mergers & Acquisitions bei Anwendung des Discounted Cash-Flow-Verfahrens (Entity Ansatz), *Die Wirtschaftsprüfung*, 621-630.

Peemöller, V., J. Meister, and C. Beckmann (2002), Der Multiplikatoransatz als eigenständiges Verfahren in der Unternehmensbewertung, *Finanz-Betrieb*, 197-209.

Penman, S. (1991), An Evaluation of Accounting Rate-of-Return, *Journal of Accounting, Auditing & Finance*, 233-255.

Penman, S. (1992), Return to Fundamentals, *Journal of Accounting, Auditing & Finance*, 465-483.

Penman, S. (1996), The Articulation of Price-Earnings Ratios and Market-to-Book Ratios and the Evaluation of Growth, *Journal of Accounting Research*, 235-259.

Penman, S. (1997), *Combining Earnings and Book Value in Equity Valuation*, Working Paper, University of California.

Penman, S. (1998), A Synthesis of Equity Valuation Techniques and the Terminal Value Calculation for the Dividend Discount Model, *Review of Accounting Studies*, 303-323.

Penman, S. (2001), *Financial Statement Analysis & Security Valuation*, Boston.

Penman, S. (2003), What Matters in Company Valuation: Earnings, Residual Earnings, Dividends? – Theory and Evidence, in: Börsig, C. and A. Coenenberg (Eds.), *Bewertung von Unternehmen*, Stuttgart, 185-204.

Penman, S. and T. Sougiannis (1998), A Comparison of Dividend, Cash Flow, and Earnings Approaches to Equity Valuation, *Contemporary Accounting Research*, 343-383.

Pesaran, M. and A. Timmermann (1995), Predictability of Stock Returns: Robustness and Economic Significance, *Journal of Finance*, 1201-1228.

Pesendorfer, W. and J. Swinkels (2000), Efficiency and Information Aggregation in Auctions, *American Economic Review*, 499-525.

Petitt, R. and P. Venkatesh (1995), Insider Trading and Long-Run Return Performance, *Financial Management*, 88-104.

Piltz, D. (2005), Die Rechtsprechung zur Unternehmensbewertung, in: Peemöller, V. (Ed.), *Praxishandbuch der Unternehmensbewertung*, 3rd edition, Herne, 779-796.

Pinches, G. (1970), The Random Walk Hypothesis and Technical Analysis, *Financial Analysts Journal*, 104-110.

Porter, M. (1980), *Competitive Strategy*, New York.

Porter, M. (1992), Capital Disadvantage: America's Failing Capital Investment System, *Harvard Business Review*, 65-82.

Porter, M. (1998), *On Competition*, Boston.

Poterba, J. and L. Summers (1988), Mean Reversion in Stock Prices: Evidence and Implications, *Journal of Financial Economics*, 27-59

Pratt, S. (2001), *The Market Approach to Valuing Businesses*, New York.

Pratt, S., R. Reilly, and R. Schweis (2000), *Valuing a Business*, 4th edition, New York.

Preinreich, G. (1938), Annual Survey of Economic Theory: The Theory of Depreciation, *Econometrica*, 219-241.

PricewaterhouseCoopers (2004), *Similarities and Differences, A Comparison of IFRS and US-GAAP*, Frankfurt/M.

Prokop, J. (2003), *Die Bewertung zukünftiger Unternehmenserfolge*, Wiesbaden.

Pruitt, S. and R. White (1988), Who Says Technical Analysis Can't Beat the Market?, *Journal of Portfolio Management*, 55-58.

Quigg, L. (1993), Empirical Testing of Real Option-Pricing Models, *The Journal of Finance*, 621-640.

Ramakrishnan, R. and J. Thomas (1992), What Matters from the Past: Market Value, Book Value, or Earnings?, *Journal of Accounting, Auditing and Finance*, 423-464.

Ramakrishnan, R. and J. Thomas (1998), Valuation of Permanent, Transitory, and Price-Irrelevant Components of Reported Earnings, *Journal of Accounting, Auditing, and Finance*, 301-336.

Ramesh, K. and R. Thiagarajan (1995), *Intertemporal Decline in Earnings Response Coefficients*, Working Paper, Northwestern University.

Reilly, F. and K. Brown (2000), *Investment Analysis and Portfolio Management*, 6th edition, New York.

Reinganum, M. (1981), Misspecification of Capital Asset Pricing: Empirical Anomalies Based on Earnings Yield and Market Values, *Journal of Financial Economics*, 19-46.

Reinganum, M. (1983), The Anomalous Stock Market Behavior of Small Firms in January: Empirical Tests for Tax-Loss Selling Effects, *Journal of Financial Economics*, 89-104.

Reinganum, M. (1992), A Revival of the Small Firm Effect, *Journal of Portfolio Management*, 55-62.

Richardson, G and S. Tinaikar (2004), Accounting Based Valuation Models: What Have We Learned?, *Accounting and Finance*, 223-255.

Richter, F. (2002), *Kapitalmarktorientierte Unternehmensbewertung*, Frankfurt/M.

Richter, F. (2003), Relativer Unternehmenswert und Einkommensteuer oder: Was ist paradox am Steuerparadoxon, in: Richter, F. (Ed.), *Kapitalgeberansprüche, Marktwertorientierung und Unternehmenswert: Festschrift für Prof. Dr. Dr. h.c. Jochen Drukarczyk zum 65. Geburtstag*, Munich, 307-329.

Richter, F. and V. Herrmann (2002), *Pricing with Performance-Controlled Multiples*, Research Excellence Series, PricewaterhouseCoopers, Amsterdam.

Richter, F. and F. Timmreck (2000), Wachstum, Betafaktoren und kontrollierte Multiples bei der Unternehmensbewertung, *M&A Review*, 277-279.

Riepe, M. (2001), The January Effect: Not Dead, but Not at All Well, *Journal of Financial Planing*, 44-47.

Roll, R. (1977), A Critique of the Asset Pricing Theory's Tests, Part I: On Past and Potential Testability of the Theory, *Journal of Financial Economics*, 129-176.

Roll, R. (1983), The Turn-of-the-Year Effect and the Return Premia of Small Firms, *Journal of Portfolio Management*, 18-28.

Roll, R. (1984), Orange Juices and Weather, *American Economic Review*, 861-880.

Rosenbaum, P. and D. Rubin (1985), Constructing a Control Group Using Multivariate Matched Sampling Methods that Incorporate the Propensity Score, *The American Statistician*, 33-38.

Ross, S. (1976), The Arbitrage Pricing Theory of Capital Asset Pricing, *Journal of Economic Theory*, 334-360.

Ross, S. (2005), *Neoclassical Finance*, Princeton.

Ross, S., R. Westerfield, and J. Jaffe (2005), *Corporate Finance*, 7th edition, New York.
Rozeff, M. (1984), Dividend Yields as Equity Risk Premiums, *Journal of Portfolio Management*, 68-75.
Rubin, D. (1977), Assignment to Treatment Group on the Basis of a Covariate, *Journal of Educational Statistics*, 1-26.
Rubinstein, M. (1974), An Aggregation Theorem for Securities Markets, *Journal of Financial Economics*, 225-244.
Rubinstein, M. (2001), Rational Markets: Yes or No? The Affirmative Case, *Financial Analysts Journal*, 15-29.
Ruhnke, K. (2003), Bedeutung des Börsenkurses bei Unternehmensevaluationen, in: Heintzen, M. and L. Kruschwitz (Eds.), *Unternehmen bewerten*, Berlin, 75-99.
Ryan, P. and G. Ryan (2002), Capital Budgeting Practices of Fortune 1000: How Have Things Changed, *Journal of Business and Management*, 355-364.
Schiereck, D. and M. Weber (1995), Zyklische und antizyklische Handelsstrategien am Deutschen Aktienmarkt, *Zeitschrift für betriebswirtschaftliche Forschung*, 3-24.
Schildbach, T. (1998), Ist die Kölner Funktionenlehre der Unternehmensbewertung durch die Discounted Cash-flow-Verfahren überholt?, in: Matschke, M. and T. Schildbach (Eds.), *Unternehmensbewertung und Wirtschaftsprüfung: Festschrift für Professor Dr. Günter Sieben zum 65. Geburtstag*, Stuttgart, 301-321.
Schmidbauer, R. (2004), Marktbewertung mithilfe von Multiplikatoren im Spiegel des Discounted-Cashflow-Ansatzes, *Betriebs-Berater*, 148-153.
Schröder, M. (2003), Der konjunkturelle Zusammenhang zwischen ausgewählten Industriestaaten, in: Schröder, M. and P. Westerheide (Eds.), *Finanzmärkte, Unternehmen und Vertrauen – Neue Wege der internationalen Konjunkturübertragung*, ZEW-Wirtschaftsanalysen, Vol. 64, Baden-Baden, 19-28.
Schultze, W. (2003), *Methoden der Unternehmensbewertung*, 2nd edition, Düsseldorf.
Schumann, J. (2005), Residualgewinnorientierte Unternehmensbewertung im Halbeinkünfteverfahren: Äquivalenz- und Transparenzaspekte, *Finanz-Betrieb*, 22-32.
Schwartz, E. (1997), The Stochastic Behaviour of Commodity Prices: Implications for Valuation and Hedging, *Journal of Finance*, 923-973.
Schwartz, R. (1993), *Reshaping the Equity Markets – A Guide for the 1990s*, Homewood.
Schwenker, B. (2001), *Synergie-Effekte Nutzen* – Printed Version of the Key Note Speech Given at the 12th Business Symposium Held at Schloss Reichartshausen in Oestrich-Winkel, Germany.
Schwetzler, B. (2003), Probleme der Multiple-Bewertung, *Finanz-Betrieb*, 79-90.
Schwetzler, B. and J. Warfsmann (2005), Enterprise-Value/Umsatz-Multiplikator und Non-Financial Multiples, in: Krolle, S., G. Schmitt, and B. Schwetzler (Eds.), *Multiplikatorverfahren in der Unternehmensbewertung*, Stuttgart, 48-57.
Seeger, H. (1998), *Insiderhandel am Deutschen Aktienmarkt*, Frankfurt/M.
Seppelfricke, P. (1999), Moderne Multiplikatorverfahren bei der Aktien- und Unternehmensbewertung, *Finanz-Betrieb*, 300-307.
Seppelfricke, P. (2003), *Aktien- und Unternehmensbewertung*, Stuttgart.
Seyhuhn, H. (1992), Why Does Aggregate Insider Trading Predict Future Stock Returns?, *Quarterly Journal of Economics*, 1303-1331.
Sharpe, W. (1964), Capital Asset Prices: A Theory of Market Equilibirum Under Conditions of Risk, *Journal of Finance*, 425-442.
Sharpe, W. (1966), Mutual Fund Performance, *Journal of Business*, 119-139.

Sheehan, T. (1994), To EVA or Not to EVA: Is That a Question?, *Journal of Applied Corporate Finance*, 84-87.

Shleifer, A. and L. Summers (1990), The Noise Trader Approach to Finance, *Journal of Economic Perspectives*, 19-33.

Shleifer, A. and R. Vishny (1986), Large Shareholders and Corporate Control, *Journal of Political Economy*, 461-488.

Shleifer, A. and R. Vishny (1997), The Limits of Arbitrage, *Journal of Finance*, 35-55.

Shukla, R. and C. Trzcinka (1994), Persistent Performance in the Mutual Fund Market: Tests with Funds and Investment Advisors, *Review of Quantitative Finance and Accounting*, 115-135.

Sieben, G. (1983), Funktionen der Bewertung ganzer Unternehmen und von Unternehmensanteilen, WISU, 533-542.

Sieben, G. and H. Maltry (2005), Der Substanzwert der Unternehmung, in: Peemöller, V. (Ed.), *Praxishandbuch der Unternehmensbewertung*, 3rd edition, Herne, 377-401.

Sloan, R. (1996), Do Stock Prices Fully Reflect Information in Accruals and Cash Flows About Future Earnings?, *The Accounting Review*, 289-315.

Sofka, W. and T. Schmidt (2004), *I Like The Way You Move: An Empirical Investigation into the Mechanisms Behind First Mover and Follower Strategies*, ZEW Discussion Paper 04-87, Mannheim.

Solnik, B. and D. McLeavy (2004), *International Investments*, 5th edition, Boston.

Solomons, D. (1965), *Divisional Performance*, Homewood, Illinois.

Stehle, R. (1991), *The Size Effect in the German Market*, Working Paper, University of Augsburg.

Stehle, R. (2004), Die Festlegung der Risikoprämie von Aktien im Rahmen der Schätzung des Wertes von börsennotierten Kapitalgesellschaften, *Die Wirtschaftsprüfung*, 906-928.

Steiner, M. and C. Bruns (2002), *Wertpapiermanagement*, 8th edition, Stuttgart.

Stickel, S. (1985), The Effect of Value Line Investment Survey Changes on Common Stock Prices, *Journal of Financial Economics*, 121-143.

Stowe, J., T. Robinson, J. Pinto, and D. McLeavey (2002), *Analysis of Equity Investments: Valuation*, Charlottesville, Virginia.

Subramanyam, K. and M. Venkatachalam (1998), *The Role of Book Value in Equity Valuation: Does the Stock Variable Merely Proxy for Relevant Past Flows?*, Research Paper, Stanford University.

Theil, H. (1971), *Principles of Econometrics*, New York.

Timmreck, C. (2004), Discounted Cash Flow – Die Bestimmung der Eigenkapitalkosten, in: Richter, F. and C. Timmreck (Eds.), *Unternehmensbewertung*, Stuttgart, 61-75.

Tinic, S. and R. West (1984), Risk and Return: January vs. the Rest of the Year, *Journal of Financial Economics*, 561-574.

Trigeorgis, L. (1988), A Conceptual Options Framework for Capital Budgeting, *Advances in Futures and Options Research*, 145-167.

Trigeorgis, L. (1996), *Real Options*, Cambridge.

Vandenberghe, V. and S. Robin (2004), Evaluating the Effectiveness of Private Education Across Countries: A Comparison of Methods, *Labour Economics*, 487-506.

Varian, H. (1992), *Microeconomic Analysis*, 3rd edition, New York.

Wagner, T. (2005), Konzeption der Multiplikatorverfahren, in: Krolle, S., G. Schmitt, and B. Schwetzler (Eds.), *Multiplikatorverfahren in der Unternehmensbewertung*, Stuttgart, 5-19.

Wagner, W., M. Jonas, W. Ballwieser, and A. Tschöppel (2004), Weiterentwicklung der Grundsätze zur Durchführung von Unternehmensbewertungen (IDW S 1), *Die Wirtschaftsprüfung*, 889-898.

Wallmeier, M. (1997), *Prognose von Aktienrenditen und -risiken mit Mehrfaktormodellen*, Bad Soden.

Wang, K., Y. Li, and J. Erickson (1997), A New Look at the Monday Effect, *Journal of Finance*, 2171-2186.

Watts, R. (1978), Systematic 'Abnormal' Returns After Quarterly Earnings Announcements, *Journal of Financial Economics*, 127-150.

Weber, M. and J. Wüstemann (2004), *Bedeutung des Börsenkurses im Rahmen der Unternehmensbewertung*, Ergebnispapier zum Symposium vom 29.01.2004, Mannheim.

Whelan, C. and R. McNamara (2004), *The Impact of Earnings Management on the Value-Relevance of Financial Statement Information*, Working Paper, Georgia College & State University.

White, G., A. Sondhi, and D. Fried (1997), *The Analysis and Use of Financial Statements*, 2nd edition, New York.

White, G., A. Sondhi, and D. Fried (2003), *The Analysis and Use of Financial Statements*, 3rd (International) edition, New York.

White, H. (1980), A Heteroscedasticity Consistent Covariance Matrix Estimator and a Direct Test for Heteroscedasticity, *Econometrica*, 817-838.

Wichels, D. (2002), *Gestaltung der Finanzmarktkommunikation mit Finanzanalysten*, Wiesbaden.

Wiehle, U., M. Diegelmann, H. Deter, P. Schömig, and M. Rolf (2004), *Unternehmensbewertung – Methoden, Rechenbeispiele, Vor- und Nachteile*, Wiesbaden.

Wiese, J. (2004), *Unternehmensbewertung mit dem Nachsteuer-CAPM?* Working Paper, Ludwig-Maximilians University Munich.

Williams, D. (2001), *Probability with Martingales*, Cambridge.

Williams, J. (1938), *The Theory of Investment Value*, Cambridge.

Wilson, G. (1987), The Incremental Information Content of the Accrual and Funds Components of Earnings After Controlling for Earnings, *Accounting Review*, 293-322.

Womack, K. (1996), Do Brokerage Analysts' Recommendations Have Investment Value?, *Journal of Finance*, 137-167.

Wooldridge, J. and C. Snow (1990), Stock Market Reaction to Strategic Investment Decision, *Strategic Management Journal*, 353-363.

Yee, K. (2000), Opportunities Knocking: Residual Income Valuation of the Adaptive Firm, Journal of Accounting, *Auditing and Finance*, 255-266.

Yee, K. (2004), Forward Versus Trailing Earnings in Equity Valuation, *Review of Accounting Studies*, 301-329.

Yee, K. (2005), Aggregation, Dividend Irrelevancy, and Earnings-Value Relations, *Contemporary Accounting Research*, 453-480.

Zarowin, P. (1990), What Determines Earnings-Price Ratios: Revisited, *Journal of* Accounting, Auditing & Finance, 439-454.

Ziegler, A., E. Eberts, M. Schröder, A. Schulz, and R. Stehle (2003), Multifaktormodelle zur Erklärung deutscher Aktienrenditen: Eine Empirische Analyse, ZEW Discussion Paper 03-45, Mannheim.

MIX
Papier aus verantwortungsvollen Quellen
Paper from responsible sources
FSC® C105338

If you have any concerns about our products,
you can contact us on
ProductSafety@springernature.com

In case Publisher is established outside the EU,
the EU authorized representative is:
Springer Nature Customer Service Center GmbH
Europaplatz 3, 69115 Heidelberg, Germany

Printed by Libri Plureos GmbH
in Hamburg, Germany